Mynedrie and the Necromancer

By

L.C Willing

Copyright

Published by

BookFuel™

Table of Contents

Chapter 1

Sea sickness was a new and unpleasant experience for Mynedrie, who had never before had to sail on the deep, blue, and outstretched sea. Its constant waving and waning made her constantly fall and perish to the ground, while the crew of this ship, which bore the name Seaworthy, told her that the waves were calm; meanwhile, in her stomach, it felt more like a storm.

It was all worth it at the end, however, as today her life's worst and greatest agonising dread was captured. Put in a specially enchanted vault with the only opening, a small hatch, to give him food and make sure he was not using one of his necromantic tricks in the cover of the vault's shadowy inside.

While she had been told that every precaution had been taken, she did not trust it as she could not help shake off the feeling of dread, anxiety, and anguish she had not felt since she was a mere apprentice under her father, specifically the day that the Greyspot infection hit her village. She lay in her bed thinking about that night, day in day out, but now was no time for contemplating; however, she had a dinner to go to, which was specially arranged to celebrate this capture, though none of the attendance, except for her, had actually aided in it, or any who had perished in the attempt.

Still, it was only polite to attend the occasion, not that she had any interest in it. What made it worse was that the captain had requested that she leave her bow and knives in her room, so as not to "disturb" the guests who were attending the festivity.

She found it ridiculous that the worst necromancer that this part of the world had seen was just a few decks right beneath them, and the captain worried that her weapons were going to make them uncomfortable.

Still, even she could not deny that in some twisted sense she wanted to go, just to get her mind finally off of Drumir. After centuries, he would finally be stopped; the countless fathers, mothers, daughters, and sons whose lives he had ended or even twisted beyond recognition would finally get their justice.

As she walked to the dining hall, she saw two of the ship's ship-maidens walk through the hallway, one was an unknown face who was carrying a plate with food, a bowl of vegetable soup with a small amount of vegetables was to be found in the cup that was practically just water, with next to it old, stale-looking bread.

The woman carrying it was youthful, pale-looking with emerald eyes and golden hair voluminous and long; she looked cheerful to her companion, who did not return the favor, for the woman next to her was Mynedrie's cousin Stria, who was slightly taller than the blonde, looking with boredom and the slightest frustration. That was until her eyes locked with her cousin. Her face turned from a stale, unchanging boredom to one of anger and contempt.

Mynedrie returned the look right back at her. Looking into

her dark, oak-brown eyes, she could see that her niece still held a grudge, even after all these years; she still believed she did no wrong.

Her colleague, who was half a head shorter than both of them, said, "I can figure that you two know each other?"

"We do indeed, Drimia. Why don't you leave us so we can catch up?" Stria answered with a faint smile that came to her while showing a dark, insulting grin to Mynedrie when her colleague gazed away from her.

She gave only a stern look on her face, refusing to show any more emotion than she had to as Stria walked closer to her, standing at roughly the same height as her. She started inspecting her long-lost niece, seeing that she had no jewels of any kind on her. No necklace, no bracelet, no ring, nothing, that could betray her aristocratic heritage that she had enjoyed, just the simple, humble brown dress with a blue piece of cloth wrapped around her thighs, of her new job.

She looked at Mynedrie with a certain contempt emanating

from her face, scanning her niece from head to toe. Noticing that she wore light leather armor, the breastplate was custom-shaped to her figure and continued till her lower ribs downward; from there she wore a leather belt, to which was attached on Mynedrie's right side the sheath of a short knife which had a sharp point at the end, making it great for stabbing her foes, handy in a small place like this, too bad for her she did not have that knife on her right now. Her shoulders and upper arms were exposed to allow her better motion and greater accuracy while shooting her bow, which she was surprised to see was not with her right now. Normally, she kept a firm hand on it like she wanted her hand to merge her left hand with the weapon if she could.

Stria knew that bow was worth a lot, for it was made by master magicians and craftsmen alike, but for Mynedrie, it was worth even more. She was there, after all, when she won it at the tournament; after all, she had tried to win it herself, but archery was not her strongest skill. "It should have been mine!" She thought to herself, feeling angry and frustrated, that of all people to win that bow, it had to be her.

Its magical wood was pearl-white in color and magical in substance, as was its string, which, unlike most bowstrings, neither had to nor could be taken off and was almost as white as the wood of the bow was, except when it was pulled, then it let out the same light as the bolt of light that it produced. Which packed a punch harder than an arrow with twice the pull of a longbow.

"Congratulations, Mynedrie, it seems you succeeded in capturing him after all," she said politely. She was, after all, still an elven nobility, even if she was disinherited by her parents and polite society as a whole.

"Thank you," Mynedrie responded with a sense of suspicion.

"I was genuinely distraught hearing the losses that were inflicted on your team. Must have been hard keeping them safe," she said with a sick smirk on her face.

"It was. It took great effort to capture him... almost as much as you not cheating," she responded.

Stria's grin turned into a dark frown. She stared for a second at her niece, who in turn did that back in kind. Stria broke the silence by saying, "Be careful around this ship, cousin, you may just fall into the ocean without realizing it!"

"Is that a threat!" Mynedrie asked, already knowing the answer,

"Oh, no! I would never threaten someone as 'great' and as 'noble' as you!" she said in a clearly fake and ironic tone before leaving her cousin in the hallway.

Mynedrie looked as Stria walked away from her, her hands tensing into fists... she would, however, take a deep breath of air before she continued on her way.

Silently, they both walked through the thin hallways of the massive ship, Stria to the specially redesigned holding cells which were on the lower levels of the stern. Just above it are the ship's toilets. While Mynedrie was going to the dining hall.

In the holding cells stood a big, dark, iron vault with a big lock on the right side of the door, in the middle of that same door was a rectangular plate which could be opened to check on the prisoner and give him food.

Beside the iron vault stood a guard heavily armoured, with a spear clutched in his right hand. He was making some idle conversation with Drimia while he had a mask on that hid his expression. It was obvious that he liked the companionship; it was better than standing alone for several hours at a time doing nothing but standing there right next to the most dangerous necromancer that exists.

Drimia and the guard noticed her arrival into the room, "Ahh, there you are, I was starting to worry," Drimia said.

She looked at the room and found it an utter mess. The guard was wearing mismatching pieces of armour, some which barely seemed to fit him. Crates from the cargo hold that was right in front of the prison cells were stacked in here; on one of these plates stood the plate with soup and stale bread which was reserved for Drumir. Stria was confused about that and asked, "Why exactly do we give him food? Is he not a powerful necromancer or something?"

"Captain's orders, he wants to make sure his prisoners are taken care of. Plus we do not know if and for how long he can go without food," Drimia responded.

Stria meanwhile walked over to the guard, inspecting his gear. It was quite a collection he had on his person. On his left lower leg he had heavy metal-looking armour while on the right it was leather. The protection of his knees was just as much disorganized. His right knee was a shiny copper knee protector; the one on his left however was made of steel and slightly more oval in shape. The armour of his upper legs was partially hidden by a cloth with the captain's insignia on it, which was a gold coloured rose on a black background. If Stria remembered correctly, it was the family crest of house Dirai, a house known for their blossoming flower fields and their massive egos. If one of theirs had somehow gotten hold of this ship, it was most likely because the family member in question wanted the honor of bringing Drumir to jail and perhaps even make some money at the same time judging from the state of the guard's armour.

"Not exactly the best equipment to use given the circumstances," Stria said.

"You're telling me. When they hired me for this job they told me I would be given high-quality gear, but these pieces are so mismatched I can barely move," the guard responded.

"And they gave you a spear in this small a room? Does the captain even know what he is doing?" Stria asked.

"Don't get me started on the captain! He is the reason why I am wearing this mismatching monstrosity. There is another armour which is semi-practical but he wants to keep it clean for when we get to Kadan's island. Just look at what I'm wearing now all because he could not be bothered to buy something more practical!"

"Shows how little faith everyone had in capturing him this

time," Stria responded.

"I know, right? I still cannot believe that he is in there, like what if..."

"AAAAHHHH!"

The guard then got interrupted by a scream coming from Drimia who had been placing the food through the little opening, when suddenly the long, thin, shrivelled like a raisin hands of the vile being inside caught her young slightly tan hands, wrapping them with the palms of his purple hued hands. Her arms had been pulled in almost sadistically slow, until they had been pulled till the elbow. She tried to pull herself out of the necromancer's grasp while a dark, maniacal, giggling tone of enjoyment rushed out of the vault. Stria and the guard rushed to help her, pulling her away from the vault and trying to loosen the grip of Drumir. The sounds coming from the vault stopped, and beside the sounds of struggle coming from his three competitors and the sound of crackling wood and the waving of the sea, it was quiet in the vault... until a high and anxious scream from Drimia ended it, "AAAAHH AH HAH, no Eamala, please no." Tears started to roll down her cheeks, which made two tiny rivers start to form on her face as she could feel something soft and slimy make its way over her fingers. In her mind, it could only be one thing. His tongue! The thought of him taking a bite out of one of her hands raced through her mind filling her with adrenaline. Until thought stopped being a thought and turned into reality for her. Feeling his teeth on her pinky. First just being there then slightly pressing, then harder and harder and harder slowly pulling up the pressure on her finger.

Between the screams of pure angst, the harsh pulling of the victim and her two would-be saviours there were several little critters, three small bugs that made their way out of the vault without any of them noticing. As they moved forward from the underside of her sleeves before making their way to the ground, hiding between all the things that lay in the ship's jail. Just as it felt like he would truly bite it off, he let go. Not out of sudden remorse but because his goal had been achieved, now all he needed to do was wait until someone took the bait.

Drimia shaking without any control was escorted out of the room with only the slow and sinister laughter behind them. When they left the room and put some good distance between them and Drumir, they were able to calm down, except for Drimia who was still shaking from the experience, "That was too close for comfort! He nearly bit your finger off!"

Drimia did not react as she was still stressed out of her mind.

"Well anyway, my shift is nearly over, you guys mind if I go?" she said while sighing of exhaustion.

She looked at Drimia and said, "Or do you want me to stay?"

Drimia quickly glanced at the guard and said, "I think I will be fine thank you,"

"Alright then," Stria said, before leaving.

She closed the door of the female personnel's sleeping room. In there were two rows of beds built three high on top of each other, each row was built against a wall, one on the right and one on the left. There were already some women sleeping in their beds, resting after a long day of work. She walked to her bunk bed which was the closest one to the floor. Out of her chest she grabbed her regular clothes and started undressing. She took off her work clothes and put on her sleeping attire.

When she first started working on the Seaworthy she felt uncomfortable having to do it where everyone was watching, but on a ship even as big as that of an elven-made one, there was only so much space that could be used, so what was not needed was not put in. Though what was considered a need differed, of course, from who you asked. This ship, the Seaworthy, was not a ship to bring safety to the masses or wealth to a nation, no it was made for wealthy nobles to show off power and wealth. To bring on board their wealthy friends and foes to show them to what they're capable of. And their current captain was no different. Stria wondered who he must have bought off to get the honor of bringing Drumir into custody.

This was after all why she chose to be on this ship even after being disowned by her parents and bribed to keep silent to make sure she would not further shame them. Not like it mattered, as word travels fast in the elven aristocracy. Of course some like to give the story a little twist here and there to make it more interesting, but even if they didn't the damage had been done. There would be no family that would let a discredited, disinherited, and dishonored woman marry into their house. Perhaps that is why they gave her that 'bribe,' maybe they still loved her at least enough to give her enough wealth to not sleep on some street beneath the trees. She always thought of how if she could just get to the right person in the right position to talk to she could elevate back to her previous status, perhaps even exceed it. For now, however, she was still stuck on this ship with these commoners that saw her as nothing more than a colleague. Granted many of them were 'entertaining,' but that was just about it. No, her dreams of wealth and status were still far away into the future, for now she just had to make do with the hay-filled bed instead of her two-person-sized one, filled with the soft feathers of the elegant and majestic golden Fulji, which was also ironically enough her house's symbol.

Stria was now too tired to think of irony and symbolism however and went to sleep covering herself with a thin blanket and closing her eyes, falling asleep on the thought that perhaps her life would change, getting what she wanted most of all: respect...

As her mind drifted away into sleep she started to dream, the type she never had before. She stood in a dark mist-filled place; there was only darkness as far as she could stare. It seemed more like a nightmare only it did not feel that way at all.

"Hello there," said a griddled voice behind her. She turned around and saw Drumir stand behind her. She prepared for combat even though in the back of her mind she knew that she probably would stand little to no chance against him without weapons.

"You want to fight me?" Drumir asked.

"I sure am." Stria responded.

"With those weapons?" Drumir inquired further.

Stria looked confused. She had no weapons with her; she stood with her fists in front of her and they had rarely had to punch something. However, when she blinked for even the briefest of seconds there they were two elven-made sabers, her preferred weapon of choice.

Drumir then spoke to Stria, "Beautiful, are they not, exactly how you wanted them. Which is why I am here to give you what you want, be it swords or something more befitting of your status. I'll give it to you."

"And what exactly are you asking for it in return? Let me

guess, be freed from your cell!" Stria responded.

Drumir started to giggle as if he enjoyed her pushback. After all, there was nothing fun if they did immediately believe you. He then disappeared from her and her surroundings changed; instead of darkness it turned into her old Holmes garden. In the garden stood a small group of people, her parents, her brothers and sister, and also many other relatives like her uncles, her aunts their children, her grandparents and great-grandparents. Stria remembered it all clearly; they had all gathered there for a competition that she had personally requested to be held. She had planned it out so well. Her family had come on her birthday, including her niece Mynedrie who had been outshining her for several months but beating her in front of the family would have shown her her place perfectly.

"Do you remember this day?" Drumir asked without showing himself.

"Yes, but not particularly fondly," Stria said, while looking angry at the young image of Mynedrie.

The memory started to change again. Stria saw that her younger self and Mynedrie's younger self stood ready to hold a competition to see who of them was the better fighter, all while the whole family would be watching.

The fight started out so well. As soon as the sign had been given that they could start, she had charged and let out a volley of slices going Mynedrie's way who was struggling to hold off Stria's aggressive assault. They were just using wooden training weapons but that would not mean that there would be no pain from it. Stria kept her barrage up impressively long, keeping her niece on the defence for several minutes with almost no break in attacks. However, slowly but surely even Stria's attacks started to become just slower and her defence would be open just long enough. Stria tried to slash her niece just like so many times before; Mynedrie however dodged it at the last second sliding to the left quickly pushing forward before Stria could even react she had been struck several times by Mynedrie's blades, giving her several points on the scoreboard that was manned by one of her uncles. Three times she had been hit, therefore Mynedrie received three points, seven more and she would win the competition. She did not want this, not in the slightest.

Mynedrie looked quite pleased knowing that she had gained the upper hand in this battle as she quickly gazed at the scoreboard. She then turned her attention quickly back to her niece, whose wrathful eyes seemed to look with hateful contempt at her. She knew that Stria did not like losing; however, the face that she gazed to her had something to it that just unsettled her to her core, she just couldn't really place why exactly.

Stria charged again to her niece to launch another set of slashes and cuts. Before she was able to, however, Mynedrie ducked and made a slide with her leg and swept Stria's legs right down from under her. Stria fell to the ground before she knew it. Mynedrie jumped onto her and made two stabbing movements with her left knife before being stopped by Stria, who grabbed Mynedrie's left arm by the wrist. She, however, had another wooden knife in her other hand and poked her with it in her side two more times before that hand as well was grabbed by Stria.

Despite her hands already holding her sabers, she was able to exert quite some force onto her niece's wrist, so much so that it started to induce pain where Stria's fingers were cramping against her arms. She did not panic, however, and instead chose to retreat by using her legs to leverage away from Stria.

Mynedrie did not show it, but she was worried, for she had never battled against someone so... weird before; there was almost no regard for her own defence, and something about her felt off. Before she could finish the thought, however, she was again charged by her opponent, who let out a furious set of attacks, even some of the ones that Mynedrie was able to block simply pushed through her defences, simply by sheer force of strength. One such a hit that Mynedrie tried to block the attack with her left knife, as the saber made contact with her knife, she could hear over the loud klinging sound a crack as she felt some pain coming up in her wrist.

In shock, she dropped the wooden knife she had in it; instead, she inspected her injured hand. As she did, however, all Stria was seeing was an enemy lowering its guard and restarting her barrage of attacks. Mynedrie dropped to the ground and covered her face with her arms.

Stria kept hitting her niece more even than what was needed, until someone behind her grabbed her by the shoulder, pushed her away, and yelled, "ENOUGH!"

She tried to lash out against the voice by hitting it with her sword, to her dismay, however, she saw that the man standing behind her had caught the attack mid-swing. Not only that, but she looked into the angry man's face, seeing two disappointed, green eyes staring right into her. That expression said more than a thousand words to her. All she could do was look back in disbelief.

It's Mynedrie's father, Dunar, a veteran mageguard for more than thirty years, almost twice as long as she was even alive. The real Stria, who was looking away with the agony she felt for this moment, was betrayed by her face's expression. As Dunar pushed her to face in front of the family, she grabbed onto a thin piece of thread hanging around her neck and lifted it, exposing the black octagonal-shaped necromantic necklace that was hanging from it, giving off a light green hue coming from in and around a skull that was carved onto it.

Stria remembered their faces, both the older and younger members of the family, all looking in disbelief, shock, and worst of all, disappointment. She thought if only she had never been exposed by Mynedrie's father... if she had never used that necklace...

"Not fondly indeed," Drumir said.

"Is this why you haunt me in my dreams?! To ridicule me?!" Stria demanded to know.

"No, no, the opposite in fact, I want to say I admire you," Drumir responded.

"Admire me?" Stria said, looking confused around the area.

Drumir then appeared in front of her, with a smile coming from his face, revealing his crooked teeth, stating, "Of course, you saw an unfair fight and decided to even the odds. After all, she was trained, raised, and even the offspring of the greatest mageguard around, even your parents could not give you that kind of thing with all their wealth!"

"While that is... true does not explain why you decided to visit me in my own mind?" Stria said,

"Oh it's simple actually, I want to help you." Drumir answered.

"Help me?" Stria asked.

A wide grin started to form on his purplely and wrinkled face, while saying, "Indeed specifically to give you what you want, status, wealth...power! I will need your help first however,"

"Let me guess, escape?" Stria said.

"You either free my helping hand from its chains, or you will live to hear your own chains of servitude rattle for the rest of your days." Drumir remarked.

"You slaughtered entire villages! Ruined the lives of hundreds of thousands of people! Not to mention the monstrosities you crafted!" Stria said with a sense of insult flowing through her heart.

Drumir remained still for a second before answering, "Eleven..."

"Excuse me?" Stria responded.

"That is how old you were when you secretly looked through the small opening of your father's office door, seeing him doing business with a man you had never met before, but his face was hard to forget, sea blue eyes, dark blond hair, flowing down till it came just beneath his shoulders, you could not help, but feel a little intimidated just looking at him and you knew your father was as well.

Even though he tried not to show any of it, but his straight and stiff posture betrayed the fact that he was in fact terrified. It would be weeks later however you would know why. As you were looking with your mother in one of your city's lovely shops, you could not help but wonder what the commotion outside was all about. You went outside to look and there in a cold, black, iron cage. In it was that very same man. He had been taken captive after he had assassinated a high ranking government official.

He tried to tell the people in court that he had been ordered by your father but the evidence pointed to a fellow rival of your father. Very clever of him I have to admit,"

"h... how do you know that I never told anyone of that?" Stria said.

"The same way your father was able to get one rival killed and one imprisoned. Because he had the power, the kind of power I want to give you now, so you may sit on a throne sipping on the finest of your most favorite beverage, instead on an old creaking chair of wood, drinking its cheap, watered down equivalent." Drumir answered, Stria contemplated what he said while he continued, "The choice is yours."

Chapter 2

Mynedrie was sitting in the dining hall of the ship at the officers' table as a guest. She was not one to talk and was just playing with her food, lost in her own thoughts, until she got ripped out of them by a question, and she hadn't really focused on what the man said in time, asking him, "Sorry, could you repeat that?"

The officer sitting just a few seats away from her, with the captain almost right in between them, repeated his question, "I asked, how did you capture Drumir?"

All the other officers looked at her with the same sense of curiosity as the one that asked the question. Feeling she had little choice in the matter, she started to explain, "Well, as with every hunt, it starts with the preparations. We knew that he had hidden himself in a cave, believing that he could peacefully continue his 'work' there. Unknown to him, we knew exactly where this cave was. So, we had plenty of time to create a plan of attack..."

As she told the story, the people sitting at the officers' table all started to look at her attentively. Mynedrie, on the other hand, stared blankly at her plate as she continued recounting what happened.

“Dunar and I went into the cave while our third and final member, Arlara, was guarding the entrance that we had booby-trapped so he could be put into his current imprisonment. For he had not taken the 'main' entrance. For I had read up on previous encounters with Drumir and knew that he was always prepared for possible attacks from the front and an escape route so he could leave before the fight was even over.”

“Attacking from the back, however, had never been tried before, so we searched for any secret entrance or dug hole he might try to escape from. Days, we walked through the dirt on that rock-covered hill to find it. The other two were almost ready to call off any further search until on the third day we found it!

Dunar I entered the cave's back entrance, just broad enough and big enough to let one person walk through, bowing forward. We were taller than Drumir, so I guess that made sense. Dunar walked in front of me. He was wearing heavy steel armor polished to a shine, even in the faintest of light. He also used a longsword which shone even more brightly than the armour he wore, weirdly enough?"

"From where did he get it?" one of the officers asked.

"From the humans, just like most other heavy weaponry. However, this was not like any other longsword; this was forged for a king long, long ago, supposedly he was very noble. I do not know how, but one day the sword ended up on the market of a city called Drune."

"I have heard of that town. Supposedly, it is very wealthy. If you go to the upper deck, you might be able to see its light tower," one officer said.

"I wonder how that king's family lost the sword and ended up in your college's possession?"

"Perhaps how most royal families lose such things: the family becomes corrupt, lazy, and foolish, and an enemy or two defeats them. Whether in court or in battle does not matter. Perhaps they were forced to sell it so they could live in some sense of dignity instead of living among the common rubble," the second in command proclaims.

"Well anyhow, this Dunar must be quite honored to be able to fight with such a weapon, is he not?" an officer asked.

"He... he perished in battle, I'm afraid," Mynedrie said with regret in her voice.

A silence befalls the officers' table. Feeling that there was an ever-growing awkwardness in the silence, one of the officers tried to change the subject, "Speaking of mighty weapons, I heard that your bow is quite unique, is it not?"

"Yes, it is made with magic embedded in its string and wood, making it able to shoot arrows of white light that are even more deadly than their normal counterpart. I got it after partaking in a bow shooting contest," Mynedrie explained.

“Where is the captain? He should have already been here by now,” one officer on the other end of the table asked.

“I don't know. Last I heard, one of the shipmaidens noticed that the door to his room had been forced open, so he went to inspect. They must have stolen something if he is gone this long,” his colleague next to him explained.

The officers kept talking about the happenings of the ship as did the crew. She, however, silently ate her food thinking about Dunar. If only you had not stayed behind in that cave. If only Drumir hadn't trapped you there.

Her thought process was interrupted by the sudden shaking of the ship and rumbling of a roaring screech. Glasses and plates fell to the ground smashing into pieces as everyone looked puzzled at what was happening, “It must be one of the Drillin! Something must be wrong with it!” The Drillin master said before storming out of the dining hall to go check on them.

Mynedrie, who had a feeling that Drumir had something to

do with this, went with him to check it out as well. They sprinted to the hallways, walking past people struggling to keep themselves from falling, or helping those that already had.

In the frontal part of the ship, where they were taken care of when they did not need to push the ship forward like a horse would a carriage, there was on the right side of them a massive creature that was going absolutely wild. Its tall, thick neck sweeping its head that was big enough to eat a grown person with ease, left and right without any concern for its own health and well-being. Together with its powerful frontal fins with which it had pushed the ship mere hours ago, it was cracking and breaking the wood surrounding it like it was nothing.

"What is wrong with it!" the Drillin master demanded to know.

"We don't know. One moment it was just calm and sleeping. Then the next it suddenly started to attack the ship itself!"

one of the crew members told him.

“Have you tried to use Tamelles taste on it?” the Drillin master asked.

“We used enough for a beast three times its size and nothing changed,” the same crew member told him.

The Drillin master looked in horror as he saw the giant, green beast pound and pound, wounding both itself and the ship. Worst yet was the fact that he made the Drillin next to it start to make restless as well. Soon, they have two of those beasts battling it out in here. With not many options left, he made a decision, “Cut the troublemaker loose!” he yelled to his staff.

“But Commander, what if he attacks the ship afterwards?” one of the staff members asked.

“It does that already at least, then it won't do so from inside out,” he explained.

The staff started to cut loose the Drillins' wooden and rope restraints, who then immediately dived, leaving the room, and for a moment it seemed like the problem had been solved until the pounding started on what sounded like the starboard part of the giant elven ship.

Mynedrie knew that could only mean one thing, and she immediately rushed to the other side of the ship, knowing that Drumir had been part of this. She ran through the hall of the ship, jumping over the dazed people in it to get to him.

As she finally reached the holding cells of the ship, she saw that she was already too late. Drumir had left his cell, the guard that was watching him and the shipmaiden that she had met while also reacquainting with her niece were both turned into undead creatures by Drumir, forced to do his every bidding.

The guard carried her unconscious niece on his shoulder, while the zombified Drimia stood there doing nothing, until Drumir noticed her standing in the doorway.

He ordered his zombie to charge directly towards the unarmed Mynedrie. Drimia was slow but what her zombie body had lost in speed, it made up in muscle strength, for her muscles would tense and pressure beyond what her previous body simply wouldn't allow. For the force a zombie would put their muscles through could be so fierce it could snap their own bones, at least before they had rotted too much.

Drimia tried to scratch and grab at Mynedrie missing at every opportunity, for she was simply too slow. Mynedrie could not be less careful than normal however as one lucky grab could mean her undoing.

Further and further she got pushed back into the cargo hold, seeing from all the way back how the zombified guard loaded her niece onto the back of the head of the Drillin, which had a green crystal on its left side buried in its flesh.

She looked at her surroundings, looking for anything she could use as a weapon, noticing a simple looking table and chair, probably used by the chief logistics officer to keep track of everything used by the ship and everything coming in and out. She launched towards it grabbing the chair and smashing it against Drimia's head, which did little against her, as the chair broke into multiple pieces, many flying across the room, others simply falling to the ground. The two pieces in her hand were broken in such a way that they had become sharp at the end.

Another attempt by the zombie to grab her commenced this attack, however Mynedrie could use it in her favor. Mynedrie jumped to the right at the last moment and pushed the stake she accidentally had made into the side of the zombie's skull, killing it forever. Falling to the ground, making only one last squeaking sigh before truly dying.

Mynedrie made her way back to the holding cells. Drumir had already left with the zombie guard and her unconscious niece before she was able to get there.

She looked out of the big hole that the corrupted Drillin had made into the ship just a meter or so away from the sea's water, way closer than what was comfortable for a ship this size. The cold sea breeze, the darkness of the night and the knowledge of Drumir's presence sent goosebumps down her skin. She crossed arms and rubbed her upper arms to keep herself warm.

She looked around but there was little to find, only the dark colored waves of the sea and the dark void of the night in the distance.

For a moment she believed she had lost track of him. As she turned away her gaze from outside and back into the ship, however, she could hear once more the screechings of the Drillin and feel the shaking of the ship. Mynedrie could only assume that he had gone to the port side of the ship and tried to make it sink.

With no second of doubt in her mind, she sprinted to her chambers to get her bow and knives, for the coming battle she would have. Once she had procured her weapons, she started sprinting through the halls. This time it was filled with people who had fallen over thanks to the third sudden shaking of the ship.

She got to her chambers, got her bow and knives, she then

immediately made her way towards the deck of the ship. As she got to the entrance hall she saw that the giant half-circled gate was closed, not locked, as it had none but closed. She knew she had to push one of the huge doors open, cursing under her breath that the ship and everything on it had to be made so sizable.

With all her might she pushed one of the doors to try and get the door that was twice her size to budge. But to no avail, no matter how hard she pushed, even as her face turned red in the attempt, it would not move as the door simply was too heavy.

She stepped away from the door and looked at the room, seeing that there was a flight of stairs that was going up to where the command centre was, which is where the captain's quarters was and where the steering wheel was which showed the crew at the Drillin compartment where they had to make the creatures steer towards.

She walked towards the windows in front of those was a

balcony. She looked outside and saw that Drumir was deliberately trying to damage the ship around the waterline so water would pour in and let it sink. All while laughing like the maniac he was.

She made sure to stand at the most port-sided balcony to give herself a clear shot. Not at Drumir for he wore an amulet that created a shield any time someone tried to attack him, making him virtually invulnerable and unable to just be taken from him as any flesh that touches it will rot as will any simple wooden tool.

The amulet was placed at the centre of his scrawny torso, buried into his flesh and letting a bright, fair, green light emit out of the amulet's stone carved eyes.

Instead, she aimed at the neck of the Drillin who was slamming the ship with its massive, now blunted and scarred head. She readied and aimed her bow, she took a deep breath, focused on her target before releasing the bolt of light her bow had conducted out of itself. As it quickly flew through the dark night, emitting a pale light that shortly illuminated a large part of the deck, where few people were looking in horror at the ginormous beast. It was a direct hit, directly in one of the Drillin's muscle groups, making it let out a massive screeching roar in agony.

Mynedrie couldn't help but let sympathy enter the mind, for the poor creature was not aware of his actions but he was aware of all the stimulation, like pain. Forced to do his new master's cruel bidding by the crystal on the left side of his head.

Nevertheless, she persisted in attacking the creature, for it was its demise or the ship. Shooting light arrow after another, some of the guards started to join her volley with their own bow and arrows. Normally when a Drillin would attack a vessel in a war, they would use arrows with special hunting tips specifically designed for the thick skin of these creatures. With the suddenness of the attack, however, the guards scrambled for the closest arrows that lay around. Arrows left and right started to rain down on Drumir, the Drillin, and on the armored zombified guard. Before Drumir chose to leave the ship's presence, getting shot from behind by the guards and Mynedrie, only stopping when they got out of range and were lost in the darkness of the night.

Mynedrie stared into the darkness, specifically where she had lost Drumir out of sight, knowing what had just been unleashed, the ship was evidence of that. With two massive holes in the huge ship, one at starboard and one slightly above water level at the port side and only one traumatized Drillin to push it, it was a miracle that most of the people onboard got out relatively unscathed.

Confusion and distress hung in the air of the ship as the people onboard tried to get back a sense of order, "How could this happen?" many onboard asked themselves, Mynedrie being one of them. In the command centre, she took a moment to think about it. She remembered that the captain had to leave for someone had broken into his quarters and the captain was the only one with the key to Drumir's cell. So something had to have happened when he went to his room.

She walked to the captain's quarters. Opening the door, she was greeted by a most unpleasant sight, the captain laying dead on the floor, throat cut with a kitchen knife, that lay a few meters away from the body at the beginning of a trail of blood. It seemed to her that the murderer tried to hide the body somewhere probably to give his or herself more time before the body was found. But gave up, perhaps the individual in question simply found the body too heavy to move, not that it was fat, such unpleasantness was only reserved for the humans on the other side of the sea. No, it was more perhaps of the excessive use of jewelry he had on that made him so hard to move.

Not something a captain would do, then again this was no true man of the sea, but actually an aristocrat using the prestige to further alleviate himself in the social hierarchy and not trained for the reality of the hard task that being in the open sea entailed, to the frustration of those that had actually trained for the task at hand.

Still, she could not help but feel bad for the man. No one deserved to go out by Drumir's doing or by one of his cowardly cultists who hid in the shadows praying to gain the power of dark, twisted creations that only the most vile would consider to worship.

With the captain gone she would need to try and find his second in command before she could decide on what to do next. She walked through the ship looking for him, asking people if they have seen him somewhere.

Her searching would lead her to the cargo hold that started to fill with water which entered through the holes that had been made by the Drillin. Taking personnel command of the repairs, standing knee deep, shouting orders left and right to close, telling those that were carrying the water out, the water with whatever could hold it, while others were exporting the cargo to somewhere more dry, or on the starboard part of the ship to make it tilt its port side slightly upwards, just enough to make it rise above the water, at least so he hoped.

"Excuse me, are you the second in command?" Mynedrie asked uncertainly.

"Yes, that I am. First officer Dirrilly," he answered, "what do you need?" he continued, while staying focused on his task at hand.

"I am Mynedrie Kianistra, we were at the same table when the ship started to rumble and..."

"HEY YOU THERE, DOUBLE TIME!" he screamed at a still inexperienced-looking sailor frozen by fear, "DO YOU WANT US TO SINK? MOVE!!" he continued.

With the young sailor man responding, "Y-ye-yes sir," before hurrying away with the can of sea water.

Mynedrie looked puzzled at the sudden interruption. "You were saying?" Dirrilly asked.

"I... uhh... came to tell you that your captain has been assassinated," Mynedrie said.

"Hm..." the second in command responded unpassionately.

"You do not seem very worried about it?" Mynedrie wondered.

For a moment he looked at her with a serious frown on his face, "If you think I'm happy about his death then you are mistaken!" he explained with intense seriousness, "He was our captain for less than two weeks, even as his second in command that is little time to get to know him. Still, he was part of the crew even if he had only bought himself into the position. I make sure his body gets taken care of properly," he continued after having calmed himself.

"Then let me go directly to the point. I need to go to the coast to catch Drumir once more!"

Mynedrie told him.

"Go after him! HA, you must be mad you barely survived your last encounter. With two colleagues to back you up, I may add," Dirrilly remarked, "Even if I had anything to spare, my answer would still be no. For all we know, he is waiting for us to send a boat to the coast just so he could attack it out of the dark and give that brainwashed beast a snack!" he added.

"Very well then, thank you for your time," Mynedrie responded.

Mynedrie walked away from the conversation with disappointment, though she did not admit that his logic made sense, not that she would tell him that but it was something to consider. The darkness of the night made it easy to sneak up on someone and attack them, especially if they were on a boat. It seemed to her that she would have to let Drumir go for now. This was not a thought she easily was able to accept, but unable to do anything against... unless.

With haste she went back to the deck of the ship to look around the ship's surroundings and at around twenty meters off the ship's port side she saw a big chunk of wood drifting on the ocean waves, where it bobbed around. It was clear that it was ripped off by the Drillin during its attack on the ship. The planks and beams that held them together were all broken and shattered, as they were ripped apart from the ship. Mynedrie climbed up on the deck's wooden railing, gloves made it challenging to keep herself stable but she managed to stay in place, she looked directly down to the water right next to the huge ship, checking if there were no pieces of wood beneath it. It was hard to tell as she only had the light of the windows right next to it to light the area for her. Once she was sure however there was nothing there, she decided to jump feet first into the water. With a loud splash, she dived into the sea, covering her entire body underneath the freezing, salty waves of the ocean, before she almost immediately swam back up.

As her head came back above the ocean floor she immediately grasped for air only for one of the ocean's many unrelenting waves to push seawater into her face. As the salty water touched her tongue she almost immediately spit it back out, leaving only a sense of disgust and thirst behind. Once she had reoriented herself she started to swim towards the drifting piece of wood, which seemed to take longer than she had anticipated. Feeling her muscles work tirelessly to fight against the water and its waves just to get to the broken piece of the ship.

When she had finally made it to the support beam with planks attached to it she climbed on top of it to escape the freezing water. On what she now used as an improvised raft she attempted to take off one of the planks, with the nails that connected it with the beam working against her. As she twisted and turned the plank around the screws to get them to budge, she was called out by one man on the ship's crewman who screamed at her, "HEY, WHAT ARE YOU DOING?"

"I AM GOING TO THE SHORE TO GET DRUMIR," Mynedrie answered back.

"ON THAT THING?!" the man asked.

"YEAH, LIEUTENANT DIRRILLY WOULD NOT LEND ME A BOAT SO I HAVE TO MAKE DO WITH WHAT I HAVE, WHY?" Mynedrie responded.

"JUST BE CAREFUL THAT YOU DO NOT STAY IN THE FREEZING WATER TOO LONG. YOUR MUSCLES CAN SUDDENLY FALL OUT AND YOU WILL DROWN," the man explained.

For a moment Mynedrie looked around, then to her wet clothes. For a moment she thought that maybe this was a bad idea, that she should just simply turn back and wait until she had returned back to any elven harbor and come up with a plan. Before she had made her decision however she considered it. She remembered that any moment that Drumir was free was a moment he could hurt or corrupt innocents. Any night he was out was a night she would spend in a nightmare.

"HEY, YOU WANT ME TO HELP YOU GET BACK UP?"

the man asked.

"NO THANK YOU, I REALLY NEED TO GO TO THE COAST," Mynedrie replied.

"WELL THEN, LET ME GIVE YOU A WORD OF ADVICE. AT THE COAST THERE ARE SOMETIMES WAVES THAT PULL YOU AWAY FROM THE SHORE. DO NOT FIGHT AGAINST THEM, INSTEAD ROW TO STARBOARD OR PORT AND TRY FROM THERE," the man advised.

"OKAY, I WILL DO THAT, THANK YOU," she said before rowing with the now released plank, towards the shores of Drune.

Chapter 3

Shivering and shaking, Mynedrie reached the coast of Drune, just outside of the city that she could locate by the light that emitted from the buildings that shone from its walls and its massive light-tower, which was bigger than any other. Only the central tower of Alti'tu'ru, capital of the elven realms, was bigger, though it was not much of a difference.

Once Mynedrie found that she was able to walk to the coast, she immediately jumped off her improvised raft and directly made her way towards dry land. The muddy sand got quickly replaced by dry sand as she continued walking forward, where she stopped for a moment to lay down and catch her breath and lay on the wet sand of the beach. She quickly felt her eyes getting heavier, so she decided to quickly continue her journey towards the city.

As she walked to the city, she couldn't help but be in awe over the massive stone walls that surrounded it. Stone as a building material was mostly reserved for the richer members of elven society or the most important of buildings. Most others, meanwhile, used wood as a building material.

As she walked, she felt a drop of water falling on her face, and then another one, and another one, until massive amounts of water started to pour down from the sky. She put on her hood and made a run for the city. She ran and she ran until she found shelter from the rain underneath the gatehouse. There she stood in front of the city's southern gate, which was made of dark brown wood reinforced with a steel framing. Above the gate was a wooden hatch, which she puzzled in her mind what that could possibly be for.

On her right, in the dark tiny tunnel underneath the gatehouse, was a door reinforced with steel bars and a metal circle in its centre with which to knock. She knocked three times with the circle.

She had to wait a moment before a guard opened a hatch above the circle and asked with boredom in his voice, “Who is there and what business do you have?”

“I am Mynedrie Kianistra, daughter of mageguard Duran Kianistra and alchemist Mayiaria Mianstra-Kianistra. I bring urgent news to your ruler!” she answered, the urge in her voice betraying her need of haste.

“Sigh, the king is a busy man, but I'm sure that if you come back in the morning, that he wants to hear you out,” the tired militiaman said.

“I do not have time for that!” she said while removing her hood from her head and taking out a silver bracelet on which was carved her family's crest: two crossing spears with two branches growing and spiraling around them, with underneath the tip of each spear a lush thick bush growing from the end of the branches. “I'm here to warn your king of a threat that had escaped our custody and entered yours!” she continued.

The man behind the door closed his hatch. Mynedrie sighed as she started to contemplate what else she could do to get inside. Then she got disturbed by the sound of the door's massive lock opening. Soon after, the door itself started to open, accompanied by loud screeching of metal against metal of the door's hinges. The guard who had almost denied her entry said, “Come in.”

She stepped forward, walking into a room which was lighted by torches hanging from the walls. On her right was a simple wooden chair; a few steps from her on her left was a door made of simple wooden planks and a small lock on it. Right next to it was another chair; on that chair was another member of the city's militia just like his older colleague. He wore a metal helmet with the shape of a brimmed hat and a leather strap dangling from each side, with on the end of the right strap there being a metal slide through which the left strap could be pulled and hold the helmet in place. Beneath it he wore a thick dark green cotton vest on which were black lines that were supposed to symbolize the mouth of the river with its island in the middle and everything. As for the armor on the arms and legs, both were made of steel plating.

“My name is Sergeant Dave, and this is Private Jake. We will escort you to his majesty's hold,” the tired, grey-haired sergeant said.

“That will not be necessary; I am sure I will be able to get there on my own,” Mynedrie responded.

“I must insist, if what you say is true, then we must escort you to ensure your safety, plus they will not let anyone in without a guard to vouch for them.”

“Very well,” she said.

Both of the men took a spear from the weapon rack and opened the door leading into the city. Before they made their way to the castle, however, Sergeant Dave shouted to the guards on top of the gatehouse, “Hey, we are off to bring her to the castle. One of you will have to guard the door until we get back.”

“No problem, will do Sergeant,” one of the guards shouted back.

Back into the cold rain, she could not help but shiver, her

teeth making a constant ticking sound which was noticed by the much younger private. "Do not worry miss, the castle is not far off from here and you will be out of the rain soon enough."

Soon enough would be now... she thought to herself, "G-good to know, private," she said as the shivering distorted her speech.

Mynedrie walked through the street with the old sergeant in front of her and the young private behind her. She stared above looking at the light shining from the top of the gargantuan lighthouse. She had seen other lighthouses on the shores of her homeland, but none were as big as this one, nor was there light as big as this one's.

She heard sounds coming from her left, seeing that a tavern had kept its window open despite the rain. No water had entered, for above the window there was a roof. There was no glass however, despite in her hometown every house and place of business having a window with glass.

They kept on walking until the king's hold was slowly revealed behind the corner of the last house on the street. Just like the city's walls, the walls of the keep were made of huge, heavy, beige stones.

Also just like the city's walls, it had a gatehouse above the gate which was underneath an arching tunnel. Unlike the gate of the city, however, the gate of the castle was open with two guards watching over the entrance. They seemed like they were laughing about something, before the sound of someone coming towards them made them take on a battle-ready posture, their halberds pointing towards her and their two comrades. The soldier on the right started to speak with as much authority as he could, "Who goes there!"

"It is me, you paranoid idiots," Dave said.

Their posture relaxed soon afterwards with the guard on his left saying, "Nice seeing you here Sergeant, nice weather we have, don't we?"

"Enough with the jokes, private, I have here an elf woman

claiming to have knowledge about a great threat that came from across the sea," Sergeant Dave explained.

The guards both looked with interest at her. Elves were not a rare sight in the trading city, though one that wanted an audience with the king was. Their sudden staring made Mynedrie feel awkward and started thinking to herself, why are they looking at me? Is there something on my face? Maybe some food from the dinner? Why does that always happen!

"So, can we enter?" Sergeant Dave asked, not even trying to hide his impatience.

"Well, you do outrank us so... yeah," the guard on the right of him said.

And so they were let through the castle's gate and entered its courtyard where within one of its corners was the lighthouse that she saw shining before, protected from outside forces by the castle's massive, beige stone walls.

On her left there was the massive keep, which also housed

the kitchen, stables, a chapel for their religion, a great hall and of course a throne room for the king, for all his decision making. It was made out of red bricks made of clay that was harvested from the river. The stones were stacked up in a row of rectangular bricks, with on top of that a layer of square bricks overlapping and then another layer of rectangular bricks overlapping above that. The mortar in between the stones was pale white and even seemed to slightly jump out of the walls, but remaining sharp and straight despite the building's age which was obviously at least a little over a hundred years old. Above the windows and doors of the keep there were stones in a grey-yellowish color, which had been built around the top like an arch.

Mynedrie walked underneath one of these inside through the keep's dark, oak double-sided door, which were beautifully carved and had shining, polished metal along the door's edges.

She walked through the doors into the building. Her soaking

boots and cloak dripping water all over the marble floor, leaving a wet trail of water behind her as she went to the king's court. Just in front of the entrance there were two militia members. Behind them were the big wooden double doors with gold finishes. On both the doors was the image of a shark carved out of it which had its jaw open towards the other shark.

They stopped when the guards crossed their spears and the soldier on her right told them, “HALT, on the orders of the king you are to wait here until he is ready!”

And so she did. She waited there for a couple of minutes in her cold wet clothes together with the four militiamen that stood waiting, or guarding. Mynedrie could feel her eyes become heavier as time passed on. Even for a moment, she started to almost doze off before waking up again thanks to almost falling over.

Then finally, after long waiting, the doors went open revealing the courtroom of the king. A long, soft, red carpet lay on the floor stretching from the entrance to the king's throne and those of his queen and heir. Mynedrie entered the throne room, her feet leaving dark stains on the carpet as she went. She couldn't help but notice the glass stained windows on the right side of the room. She had not seen them from outside however, the view had been blocked by wooden shutters that had the city's insignia painted on them.

Mynedrie found them beautiful but struggled to get any meaning out of them. One was the mouth of the river Dru where the city now is. Then came an image of three men: one worked at a forge with metal, one worked at a workbench with wood and last was a man who had captured many fish with a net. Then came the image of the lighthouse that seemed to shine its light into the darkness and last was the image of a well dressed king, with a knight's helmet in his right hand and a sheathed sword in the other.

She now stood just a few meters from the king himself who was a tall man, taller than average that much was clear. His face was partly covered by a long but well trimmed beard and mustache. The hair on his head was black but showed signs that it was starting to get grey. The crown that rested on his greying hair was a fair golden crown with jewels shining from it. Underneath his face he wore expensive clothing made of the finest fabrics. On top of these clothes he wore jewelry of gold and jewels of many varying colors. Mynedrie made a long and deep bow to show respect to the king, before she could state her case towards him, "Welcome, my guest. I heard that you bring a warning for me in this late hour," the king greeted her.

"I thank you, your majesty, for meeting with me. I came here with quick haste to warn you that Drumir has escaped imprisonment!"

There were a few seconds of silence before the king reacted, "Drumir, you said?"

"Yes, your majesty."

"Who is now on our side of the sea?"

"That is correct, your highness."

"You said that your people had been able to capture him, do you happen to know how?" the king asked.

"I do indeed, your majesty, in fact I was the one that led the mission," Mynedrie explained.

The king's demeanor seemed to calm hearing this, telling her, "Good, good. That means we have your experience to draw from, miss?"

"Kianistra, your majesty."

"Well, miss Kianistra, I will have my city guard make preparation to find and if possible destroy that mad magician once and for all, with the aid of your input of course."

"But of course, your majesty."

"Well then let me provide you with a room so you can prepare and... dry yourself, we cannot have you getting sick, can we," the king jokingly remarked.

"Thank you, your grace, I would appreciate that immensely," Mynedrie said, before she made again a long and deep bow before turning to leave the room, with the guards escorting her to her room.

As she walked through the keep's hallways she could not help but think how humiliating it was to stand before a monarch with water dripping all over their expensive carpet. The king showed not to mind however, probably understanding what the real threat was to the city. Anyway, she was glad that she had been given a resting place by the king. Her muscles were sore, her skin cold and her body exhausted.

“Hey, private, go and find the chief of staff and tell him we need access to one of the guestrooms,” the sergeant ordered.

“Yes, sir,” the private answered.

For a while they waited before the private returned with a maid, who was an overweight, short woman that seemed to be in a good mood even with the time of day and bad weather outside. The woman opened the door and gave the key to Mynedrie.

“Thank you,” she said. Mynedrie entered the room that she got to use as long as her stay in the castle would take. It was quite large. It had a nicely made up bed, which was placed in the far left corner of the room with a bedside table next to it. Right next to her was a door. Curious, she opened it. Behind the door was a bathroom. Most of it was familiar. There was a bathtub placed against the wall right in front of her. What struck her odd however was the weird curved pipe that stuck out of the wall with a button on each side of it that seemed like the steering wheel of a ship. They seemed similar to each other, except for the fact that the one on the left was coloured blue in the center while the one on the right was coloured red. As she moved forward to inspect the odd pipe in the wall, she was suddenly surprised by an unexpected sight on her right... it was a toilet. In the same accommodations she was in. For a moment she felt sick to her stomach but was soon distracted from her disgust by the large box which was slightly bigger than the torso of a man stuck in the wall above the toilet, with rope protruding from the left of it.

Mynedrie walked back to the front door of her accommodations to ask the maid what it all was, but she had already left to do some of the laundry or something like that. Which brought her back to the mysterious rope and wheels that were in her bathroom. She first stared at the thin, dangling rope and carefully tugged on it which made a sudden influx of water appear and almost immediately disappear again through a hole in the toilet.

Mynedrie took a step back in bewilderment and said under her breath, "Magic!" She moved back to it while keeping her guard focused on both the toilet and the box above it, which made a peculiar hissing noise. She pulled on the rope again and once more water started flowing into the toilet, "But harmless," she said feeling calm once more.

Next up, she moved to the wheels at the bathtub. She turned the red wheel first and a short moment later water came pouring out of the tube. She reached out her hand to touch the stream of water. It felt cold to the touch, but quickly she noticed that it was getting hotter and hotter. She wisely pulled away her hand before it could get burned. The sudden increase of heat came as a shock to her but it was now rather obvious what the colors on the wheels meant: red is for warm water and blue is undoubtedly for cold.

The idea that water, which you did not need to get from a river or well and even warm if you wanted it to, this sounded like master levels of magic only she had not seen such magic even from the best of mages, especially not on a citywide scale. It honestly took her a moment to process such a feat, staring for a few seconds towards the tap in awe... then a thought occurred to her, if I do not need to get out for water I can stay in a bathtub until I am as wrinkly as an old human woman.

She turned the cold water on as well and decreased and increased on both of the wheels until the water that flowed out of the pipe in the wall reached perfect temperature.

As the water slowly filled the tub she looked around the room. Right from her was a table, next to the wall at the table were two chairs, across from that and next to the bed was a window from which you had a view from the bailey and of course the big light tower which was standing in its own big separate corner of the castle.

It almost painstakingly stood out from the rest of its surroundings just by its sheer size, it being at least ten times larger than the surrounding towers of the castle and city walls. She had heard about it of course, but actually seeing the behemoth of a tower was something else entirely.

Mynedrie moved away from the window and back towards the table on which she lay her pouch-belt, in which sat all her medicine, some food, a wetstone and some rope, all things a mageguard needs to survive. She then took off her leather arm braces that protected her arms both against her enemies' attacks and the string of her bow, she then did the same for the armor on her knees and lower legs.

She walked back to the bathtub and placed her weapons right next to it so she was ready for any attack coming her way, before she was going to get into a well deserved bath.

Chapter 4

Mynedrie lay in bed, however, she did not sleep easy, for she did not dream peacefully. Instead, she was currently surviving a nightmare—the same nightmare she had to survive for over ten years.

It always started out the same: she walked with her father towards the town's stables to get on some horses to patrol the nearby swamps. Little did they know that it was something from the swamps that would come to them that day.

Always, it was there in the dream—the same place, same moment—and every time Mynedrie met it again and saw its thin, malformed, pale, disproportionate body her nightmare would end.

Waking up with a dry sweat and fear, despite it being long gone. To calm herself, she walked to the tap in the bathroom, formed her hands in the shape of a cup, filled it with water, and splashed it on her face. The cold water woke her up entirely and also washed away any sweat from her face.

She took a good look at herself in the mirror. Underneath her light-blue eyes were darker-blue circles. Her brown hair, that over the day was braided into a knot, was now loosely hanging in front of and behind her shoulders. She let out a tired sigh as she looked in the mirror, telling herself, "He is gone, he is gone, he is gone..."

As she finished talking to herself, she noticed that some of the water had fallen on her new shirt that she had been given by one of the maidens of the keep. It were just a few drops of water around her neck collar, nothing more. It was frustrating enough, however, for her to let out a second sigh. She left the bathroom and walked to the window that was covered in raindrops, some that slowly traveled downwards on the glass. She was focusing on what was happening behind them, however, there was no one outside except for two guards on one of the defense towers of the castle, which was armed with a ballista. She then turned her attention to the lighthouse, which was still shining as brightly as when she just came here; it was truly much bigger than any other. Just as she left the window to return to her bed to try and sleep again, she felt the room shaking, and then again harder this time, and then another time even harder, as if something was coming closer. Then came a loud screeching roar, which she recognized as coming from the corrupted Dirrillin. She moved to the window to look if there was anything to see, and to her shock, there was, as she saw a dark figure moving just behind the lighthouse. It was attacking the city walls and militia stationed there.

With no further thinking, she immediately grabbed her stuff: her bow, her knives, and her pouch belt, and moved to get out of the room. As she turned, the door open, she noticed one of the guards just calmly moving through the halls away from her. She yelled to ask him, “Hey! Where is your king? I need to speak to him about the Dirrillin that is attacking the walls!”

The guard said nothing, only slowly turning around,

revealing his sickly face. On his forehead was a small, rough, purple crystal, which had turned his skin grey with purple lines running all over it, with the white in his eyes turning to have a purple hue on them. His face was expressing anger and with a massive, wrathful screech, the corrupted guardsman launched his sword at her attempting to stab her with the long weapon while protecting itself with its shield.

As Mynedrie dodged his attack, she noticed that the fighting stance was there, but it was off, as if it was imitating a usual soldier that was fighting. As she easily dodged one of his sword slashes, it buried itself deep into a wooden table, cleanly cutting off one of the table's legs. Mynedrie knew that that crystal was Drumir's doing. It seemed to take control of people, making them less skilled, but far stronger as well.

She dodged another hit by the sword and quickly countered with her knives stabbing the soldier in the wrist and chest. But as one knife tried to pierce the leather armor around the wrist, it was deflected as the curved knives simply shoved off it, same with the chainmail that was underneath the cotton vest of the corrupted guard's armor. The guard tried to hit her with his shield, but again she dodged it. She found herself in a disadvantageous position. There was little moving space in the hallway and she could not get enough distance between her and the guard to get her bow and shoot him with that instead.

It then dawned on her that perhaps she could still kill him from a distance. She turned her knife in her left hand, holding it at the tip of the blade rather than the handle, and used the right one to aim. Then, just as he prepared to slash her with his sword, she threw the knife towards his face. It was a direct hit, killing him instantly, making him drop to the ground lifeless.

Mynedrie walked towards the guard's lifeless body looking at it sad, knowing that the man had no choice and was turned without any say in the matter. She took her knife back and looked at the blood that was on it, contemplating if she had killed an enemy or an innocent man.

Before she could find an answer, however, she heard the footsteps of more soldiers approaching her position and sure enough, two corrupted militiamen came around the corner both carrying massive spears with them.

Seeing no other way to get to them, she threw a knife at each of them. Both knives launched into their faces and made them drop dead on the floor. Before she could go and get her knives back, however, two groups of corrupted guards came running towards her from both sides of the hallway.

Mynedrie retreated towards her room, barricading the door with the bed.

She opened the window and climbed upward onto the roof of the keep. On the roof, she was able to get a good look at the situation. She saw a small number of soldiers fighting against their corrupted brethren, and in the distance, she saw the Drillin wrecking the city defenses. She knew that Drumir was close, but that beast was the greater problem now.

She started running over the roof towards the castle walls, the roof tiles clacking as she stepped on them while keeping her balance on the sideways roof. On the ground she saw an entrance to the top of the wall. In front of which was a corrupted guard member. Mynedrie took her bow, aimed it at the guard that was in her way, and shot him right in the chest. The light-arrow disappearing soon after it came to a halt. She climbed downwards using a rainpipe, which was placed at one of the keep's back corners.

She moved up the rotating stairs, coming on top of the walls. She turned right, running closer towards the creature. She came across the first guard tower in her way, which had an iron gate in front of its dark wooden door.

The gate and door were both locked. Mynedrie slammed at them while also screaming, "Hey, HEY, DO YOU HEAR ME? I NEED TO GET THROUGH THIS TOWER!"

"HEY, WHAT ARE YOU DOING?!" one of the guards

asked, who was busy on top of the tower, helping with the ballista, that was shooting towards the creature. However, at their range, the big bolts of the ballista seemed to do little.

Mynedrie looked above toward the guard that was looking down at her and said, “I NEED TO GET THROUGH YOUR TOWER!”

“WHAT? ARE YOU CRAZY, IT IS A BATTLEFIELD OUT HERE!”

“I HAVE A PLAN TO KILL THE BEAST!”

The mustached guardsman looked at his colleague who was busy trying to reload the ballista on his own, telling him, “She says she can kill the beast.”

“Well, if she thinks she can, I say let her, cause these things sure ain't doing the job!” he replied.

He looked back down and screamed towards her, “OK, WAIT THERE FOR JUST A MOMENT.”

And so she did until the gate and door went open. She stormed inside to continue her path, while the guard unlocked the locks of the door and gate on the other side and asked her, "So, what exactly is this plan of yours?"

"Well, my good sir, you are about to find out."

She ran towards the creature, watching out not to step on any of the ripped apart bodies on the floor, their blood soaking the stone tiles of the city walls.

She ran to the tower which lay closest to the harbor. Right next to it was the outer chain house, a reinforced structure from which the outer iron chain could be lifted and lowered to prevent ships from sailing in or out of the river, or trap them in the harbor.

The gate and door of the tower were wide open. Whether no one decided to go to the ballista or was scooped up while attempting to do so was hard to tell. It was of no concern for her, however, as she was going up to fight the beast itself. She walked up the stairs of the tower with great haste, all while hearing the loud muffled screams of men and the screeches of a maddened beast on a rampage through the circular walls.

On the top of the tower she found a half-loaded ballista and the lower half of a soldier whose entire upper half must have been consumed by the Dirrillin while she was held up by the tower just before. She examined the ballista considering if she should maybe try to reload it. In the end she did not, as she did not know what the different parts of the artillery were supposed to do.

She shot normal bolts towards the creature, which was taking the ballista's bolts and the militia's arrows without noticing how they were launched into its undead, scaly skin. The magical white arrows of Heartnailer seemed to hit differently, however, as it immediately turned its attention from the guard towards her. It immediately moved towards Mynedrie and climbed upwards to the top of the tower, using the long, slimy, scaly fingers with claws that had grown on the upper parts of its frontal fins.

As it climbed upwards, Mynedrie stood still, not moving,

with her bow drawn all the way, staying concentrated at the spot where she expected the creature to rise headfirst above the tower.

She had to stand there to build up the energy to turn a normal light arrow into an explosive arrow.

As she continued waiting, she felt her arms shaking as the pressure of keeping her bow ready to fire started to take its toll on her muscles. Still, she remained focused on the beast's expected point of rising, hearing its massive claws digging itself in the stone and soon after leaving the walls of the tower, pushing itself further and further upwards to get to her. What must have looked impressively fast from the militia's point of view was seeming to take hours for her. She could feel her arms grow more and more restless wanting to let go of the force that she had been pulling for the bow.

She then saw the image she was waiting for. Both a sense of relief and fear washed over her as she saw the head of the Dirrillin come out from underneath the tower.

Just as the creature moved slightly back to prepare to launch its knife-filled mouth at her, she let go of the string of her bow, making the arrow fly through the sky. As she let go of the string, one thought raced through her mind, I better not miss. Sure enough, she did not, as the arrow made contact with the creature right under its head. The explosion caused the monster's neck to rip apart and with that its head to sever from the rest of the body.

The monster's body fell fore over on the tower, causing it to partially break apart with Mynedrie falling almost two meters downwards where her fall was stopped by a section of the circulating stairs of the tower.

She fell hard on the steep stairs even gliding a few staircases downward before debris on the stairs stopped her. For a few seconds, she lay there on the stair, motionless, believing she had died. She only got up and looked at her surroundings after she was certain that was not the case.

The surroundings were unclear at first, for dust was all

around her. But then she noticed the body of the corrupted Dirrillin, which lay motionless next to the ruined tower. When the dust had settled enough for the rest of the soldiers, they could see how the beast's body lay dead next to what was once a defensive tower. While the soldiers cheered that she had defeated the big beast, she did not. For she knew that this rampage was not his decision but forced on the creature by one of Drumir's cruel and uncaring crystals.

She had no time to think about it, however, as Drumir's corrupted guards moved in to attack the few defenders that were still around. From the tower, she saw how the mere ten counting members of the city's militia were overrun by the nearly twenty-five counting corrupted militiamen. It was a slaughter. Despite their desperate counter, they were simply overpowered by the possessed added strength and endurance, in the chaotic mess that the militia had been placed in.

Through all of that, most were slain in battle; those who were left would be taken captive, which seemed to Mynedrie what they were ordered to do, and it was almost certain that they would be corrupted by them as well.

She considered shooting them from her elevated position, but saw that several of the corrupted guards were carrying crossbows, which they would be more than happy to use against her. She had the advantage that she was not yet noticed by them, and so she could carefully try to sneak away from the tower towards a safer position where she could rest and make plans for her next move.

She was close at the entrance of the tower. She looked back at the soldier who had been captured. She could swear that she saw one of the soldiers looking at her, as he looked at him, she felt a sense of shame growing in her. A voice in her screamed to shoot the corrupted guards and save them, but she was injured, the fall had bruised her severely, and she missed much of her equipment: her knives, her armor; she was in no condition to fight anyone.

She turned away from the soldiers with every step torturing

her soul even more. Once some distance had been placed between her and the battlefield she had just escaped, she could feel her body calming from the rush of battle. This did not bode well, however, for as the adrenaline in her body dropped, so too did it grow more weary and weak. She felt how her legs became heavier and heavier to move and her head dizzier as she tried to continue moving forward.

Unable to go any further, she went to sit next to a bunch of wooden crates. She found herself dangling on the edge of unconsciousness. Her dizzy mind makes everything move in front of her and unclear to see. Despite the uncertainty that her own mind subjected her to, she could make out one figure in the distance of her dancing surroundings, which was in the shape of a man. Could it be a corrupted guardsman? Is he here to take me to Drumir? She thought to herself, if so, she could do nothing to stop him from doing so, as all her energy had been spent and exhausted. All she could do now was watch as the mysterious figure came closer and closer to her, before she truly lost her consciousness and fell to the cold, wet, hard, stone-paved ground.

Chapter 5

Stria woke up feeling a cold breeze running over her. As she opened her eyes, she was greeted by the bright shining stars of the night. They seemed to greet her happily, yet cold at the same time. She tried to cross her arms to warm herself, however, she quickly found that she was held back from doing so, for her arms had been tied up as well on a table, as well as her legs.

She looked around to see where she was. She saw on her left a huge pile of burned charcoal and ash-turned wood which was placed underneath a roof, which in turn was held up by a wall on one half and a series of thin bronze pillars, all of which had rusted to be green in color.

On her right was a bronze-made rail, behind which was one big shade of darkness, only interrupted by the stars, which shone their light in the night.

While it was too dark to notice it, she knew she was close to the sea. The sound of the waves, the cold breeze in the air, even the subtle smell of salt water revealed what was behind the darkness. After years of working on the sea as a ship maiden, she seemed to have gotten a sense for it. Which meant she was at the top of a lighthouse.

Then, out of nowhere, she heard a giggle coming from behind her, from her left. Drumir appeared with a sickly grin; he gazed at her. "What are you doing?! Why have you tied me up like this!?" she demanded to know.

"Is it not obvious? I am helping you," he answered.

"Helping me!"

"Yes, you wanted me to help you become stronger—strong enough to beat your niece, to make her bend to your feet and acknowledge your superiority over her, with a whole nation right behind her doing the same!" Drumir explained.

Stria seemed to picture it in her head: a whole nation in her control. She had not forgotten her current situation, however, so she asked him, "By tying me up on a table on top of a lighthouse?!"

"Oh, do not worry, this will only be small changes, not more than a few minutes, after which you shall fight the remaining forces of the former regime, right by my side, and after that you shall rule this land as a queen!" Drumir proclaimed.

"Under your supervision, I assume," she said.

"No, no, not at all. I am not interested in the power of government and ruling from a throne room. I want to create things that keep kings and queens awake at night, considering and contemplating how they will rid their lands from my beasts. All I will ask of you is that if I happen to come into your realm, you will look the other way when I take some... 'resources' with me if I happen to roll into town," he said, with a huge grin on his face.

Both curiosity and fear rushed through her system. She wanted to become stronger and gain the status she knew was hers by right! But did she really want it to be through something like this? Then again, what did she have to lose...

"Tell me, what are you going to do exactly?" she asked, the

shaking in her voice betraying her worries.

Drumir showed something to her that he had in his hand: two well polished, grey oval-shaped stones, with rigid crack-like shapes from which emanated a dark yellow light. They were quite the sight to look at; one was noticeably bigger and more rounded in shape than its smaller, thinner counterpart.

He grabbed from out under his robe a glass container, in which was a grey liquid. He placed it on the table on which Mynedrie was tied up. She looked curious and confused at the container, wondering what he must have put in there.

Lastly, he took out a knife and moved closer to Stria who, understandably concerned, asked him, "W-what do you think you're doing!?"

"Relax, you think I would go through all the effort of having you tied up by my new lackeys just to kill you with a knife? You have not spent much time studying my work, have you?"

"Do I need to?!"

"Not really, but you will need to sit still for a moment,

understood."

For a moment, she looked Drumir in his yellow eyes, whose whites were shaded in green, pausing to say a word to him, knowing that her answer would have dire consequences. "Yes," she answered, knowing that from here on it would mean a most serious endeavour would be undertaken by her. He placed the smaller stone on the middle of her forehead, just above her eyebrows. When it touched her skin she felt an aching, burning sensation where the thing had been placed, like it buried itself into her skull. He then grabbed the fabric just under her neck, ripping apart her work clothes with his knife till just above her cleavage and then placed the bigger stone right above it. Again, she could feel it burning and melting through her skin until it, this time, reached her ribs.

He moved to get the glass vat with the grey substance. He opened the can and started to pour it all over her body. Almost as soon as it touched her skin, it got absorbed by it spreading over her body, turning her skin from fair and slightly tanned into light-grey and shining. It was not just her skin, however, as her short, walnut colored hair, her leaf-green eyes, and pink self-manicured nails also turned into cold unfeeling grey.

Despite all the liquid having already been absorbed by her skin, she could still feel it moving like an angry liquid that was flowing and splashing into and onto different directions, trying to find its way through her body. Painful it was not; annoying it was, with frustration hitting soon after. “What is that stuff doing to me!?” Stria's voice squeaking with fear, demanded to know.

“I'll tell you when it is finished,” Drumir answered. He then took out a small pouch in which was a bottle. In that bottle was a dark, troubled, puke-green liquid which was held inside the bottle with a simple cork squeezed on top of it.

Drumir walked to the edge of the lighthouse balcony. There he pulled out the cork and let just a little bit of the liquid touch his right hand, around which grew a dark, swirling fog, which hid his hand behind it. Then immediately after, some of the fog started to rise upwards, towards the sky where it would culminate into a huge, mighty, and thunderous cloud. Among the other rain clouds it stood out for it was thundering and flashing with green light. For being such a mighty cloud it hung abnormally low only around twenty meters above the lighthouse.

Stria pulled her head forward as far as she tried but found it nearly impossible to do so for whatever Drumir had put over her and got absorbed into her body, it was heavy. Despite this added constraint she was able to move her neck to see Drumir walk close to her while putting the bottle, with its cork back on, back behind his cloak.

He then immediately pulled out something else, a seed and a spark of purple lightning directed at the little seed. He then crouched down and placed the little seedling in between two of the wet tiles of the lighthouse. Drumir walked back towards Stria who found it harder and harder to breathe with her body becoming more and more metallic, with her muscles straining under the sheer strength they needed to exude, making it feel like her chest was being pressed by the weight of an overweight man that lay on top of her. "Enjoy the show!" Drumir said smirking.

Almost as soon as he said that a massive branch started growing but something was off about it. It was bare; no flower or leaf grew on the massive root, which was a sickly shade of green, with some parts being black or covered in white powder, like it was already starting to decompose.

With the wave of his hand a sub-branch started to move towards them. When it reached them it grabbed onto the table Stria was stuck on and lifted her upwards until it reached above the lighthouse.

Stria looked at the dark thunderous clouds which roared right at her face with ear-deafening volume. She looked with angst as she realised what was about to come her way. Doubts were starting to plague her mind: was Drumir truly helping her to become stronger and regain at least some of her lost status? Or was he just simply trying to have a laugh at her expense with some very over-the-top execution, without his victim even knowing it was happening.

Drumir looked from the balcony at Stria's now elevated table. He brought his point finger, middle finger and his thumb of his nearly bone and skin and fog covered hand ready to snap them, “Here we go.” he said before acting. As soon as he snapped his fingers a bright flash and loud thunder came from his cloud, unleashing a enormous bolt of lightning onto Stria. The bolt hit her so fast she could not even scream, not that she could have been heard over the green lightning strike's enormous thundering.

While the lightning flash lasted less than a moment, it clearly had a major impact on the branch and the table it had carried upwards, scorching them what seemed immediate and leaving them burning in massive, wild flames, that even the heavy rain seemed to have trouble with extinguishing.

He again waved his hand to make the branch put down the table, which was still burning together with the upper half of the rotten branch. Creaking and screeching it moved looking as if it would almost fall apart on the spot. Lucky enough, however, it did its job and placed the table where it had picked it up in the first place.

Drumir walked to the largely scourged table, with Stria still laying on it practically harmless, except for her scourged clothes, though lifeless in appearance except for a faint breathing. Drumir checked if she was truly alive, getting closer to hear her breathing, then checking if her blood was still flowing by putting two of his fingers against her neck.

“Good, good, she is alive. I guess she worked as the main ingredient after all!” he then touched the large stone on her chest, now shining with green energy through its yellow lines, on her chest, and immediately after the one on her head, which woke her up with a huge gasp for air.

Stria was in shock and moved away from the burned wooden table, falling to the ground in the attempt. As she tried to get back up she saw that her arm was still metallic grey, but she no longer felt it moving in her body. She then noticed Drumir, who looked at her with perfected fascination, "What have you done to me?!" she demanded to know.

"Exactly what you asked for, I made you stronger. All it took were some... changes!" he said as he giggled.

As she looked away trying to take it all in, she realised that it was no time for doubts. After all, Drumir promised to make her a queen and to be a queen meant you needed to make some harsh decisions, did it not?

"Would you like a mirror?" Drumir asked her as he took out a small hand mirror from out of his cloak.

Curious to see what damage he may have done to her, she accepted the mirror. She nervously brought it towards her face. Fearing what she may see, to her relieved surprise it was not that bad. Yes, all colors and shades had left her face and hair (except for the white in her eyes) and except for the shining oval-shaped stone in the middle of her forehead, it was overall still the same.

Having calmed down after her initial shock, she decided to stand up. She felt still like herself, yet also far more energetic than before. She carefully took her first steps. Walking did not feel any different than before.

“Try to punch that support beam there,” Drumir asked.

His request was odd, but she knew that he was many things; bad at his craft was not one of them. So she walked to one of the thin, bronze pillars that supported the roof above them. She stood ready to punch it. She quickly and furiously punched with her right hand and before she knew it, she had punched a large, deep dent into it, which was roughly in the shape of her fist.

“Not bad, not bad at all, now let's see you do that without holding back,” Drumir pleasantly requested.

Stria did as he asked, wondering to herself just how much

stronger she had become. With full force, she hit the bronze pillar so hard that its upper end got ripped apart from the roof, and the under half of the pillar bent into a nearly square angle. The power was enormous, yet she felt no form of punishment on her hand. As if it had been turned into a weapon of war. She looked at her upper hand and noticed that her nails had become grey as well, just like the rest of her hand. Despite her fingertips having been changed as well in her transformation, she could still feel with them, somehow better and even more detailed in fact. With her fingertips of her right hand, she rubbed over her left hand and then the other way around. They felt cold, like steel.

"I always put my attention into the details," Drumir said.

"What do you mean?"

"Your nails. I got the idea when a cat tried to scratch me, while I was, hehehe, having fun with the village residents!"

Stria knew better than asking him about that for she knew enough about him to know what his 'fun' entailed. She instead focused on the powers she had been given, asking Drumir, “So what else can I do?”

“I'm so glad you asked,” Drumir said with a grin going from edge to edge on his face. He walked up to her and pointed towards the big stone on her chest, “That rock right there is what powers up your body. It is what gives you three times your normal strength, almost endless endurance and greater agility,” he explained, “and the one on your head strengthens your mind, giving you sharper senses, faster reflexes and even improved memory,” he continued.

“And what about my skin... my eyes and hair? Why are they now grey?” Stria asked.

“Ah, you see, the grey goo that I poured over you so nicely was a liquid metal that I had developed. I was looking for someone to test it on, before I got captured by your oh so persistent niece and her little rabble of elves. Oh well... at least I got to try it out at the end,” Drumir started to ramble.

Stria interrupted, “You used me as some kind of test subject!

Just what makes you think that you can use an untested..."

Drumir, with his voice filled with wrath, yelled at Stria, "Do not interrupt me!"

The sudden outburst startled her enough to take a step back from him. With her mind thinking he is about to attack her with one of his horrific and twisted creations. To her relief, however, he quickly calmed down and continued explaining, "But yes, you are indeed one of my little experiments, which, as you should be able to tell, succeeded. Which I was certain of. The purpose of the experiment was to turn your skin into armor. Just as strong and durable as plate armor. Your bones have been reinforced as well. If something would, however unlikely, be able to wound you, you would quickly heal thanks to the stones installed and provided by me."

With these new strengths Mynedrie will not stand a chance against me... Stria thought to herself, no one would stand against me! Her thought continued.

"Now come with me downstairs to the keep where you can

get some equipment for your upcoming mission," Drumir said.

When they made it down to the castle's courtyard, she was shocked by the sight when she saw that the militia stationed at the castle were looking odd, with dark green crystals on their foreheads. "What have you done to them?" Stria asked.

"I ensured their loyalty," Drumir responded, "Do not worry once the city is fully in our control and I am on my way to play somewhere else, you get to decide what happens to them," he explained further.

She felt bad for the human guards that had been brainwashed, even though mankind was not as refined as the elves and only lived about a third as long. To see them like this was regrettable, perhaps even worse by the fact that they were only living so short, making every moment not free even more of a waste. Still, if she became their queen, she would be able to free them from their brainwashed bondage and even improve their lives or at least get them to act in a more dignified manner, at least semi-reasonable.

As she was considering the possibilities of her reign here, Drumir was seeing something from the corner of his eye. It was a particularly large rat that walked along the castle walls and a sudden sense of an idea came brewing out of his head.

"You two grab that vermin right there!" he told two of the corrupted guards closest to the rat's vicinity.

The two guards chased after the rat, which almost escaped through a thin sewer drain, with one of the two guards grabbing it just in time and bringing it to Drumir. As they walked to him the rat was moving and biting into the guard's hands, trying to escape from his grasp.

"Yes, this one will do nicely. Send it to my workplace and put it in a cage so I can work on it later," Drumir told the guards.

They then started to walk towards the massive tent that had been built in the courtyard and took up around a third of its total area. With the rat still attempting to flee the guard's grasp.

The tent was square in form, and its sheet was colored in dark brown and yellow in color. It stretched out from one end of the castle's walls all the way to the lighthouse, with three points on top of it created by three pillars inside of it to prevent the sheet from landing and covering everything inside the tent.

“Now, my loyal lieutenant, go and get yourself some new weapons and a new uniform as well. I meanwhile have some work to do,” Drumir told her.

As they parted ways, she quickly looked back to see Drumir enter the tent with no one guarding the entrance, which was very sloppy, or perhaps it was intentional—after all, who would even try to attack the most notorious necromancer of the elven realms, surrounded by brainwashed soldiers, incapable of refusing his commands? Why she had not been given one of those crystals, instead of these stones, she did not know.

As she walked through the halls of the castle, she saw one of the maidens that had been given one of the crystals. In some twisted sense, she was cleaning the hallway with a broom. The way she did it was wrong, though—less like she was actually trying to clean the floor and more acting like the caricature of a maid and her work, like a child imitating a parent. Perhaps, there lay my answer, she thought to herself.

She got to the castle's armory. There lay many varieties of weapons: long spears, halberds, and even lances with metal tips used by knights. Then there were the swords, both long and short in length and in varying shapes. There was also quite a substantial collection of maces and warhammers, which could, with sharp spikes, tear through flesh and penetrate armor upon impact. Then there were the bows and crossbows and the bolts and arrows that were needed to fire with them.

The crossbows were a weird but equally fascinating design. It had not much popularity in the land of the elves, for the time to make one took too long for most craftsmen, who preferred to work on traditional bows. They also shot fewer arrows, or 'bolts' if she remembered the name correctly. Still, they were great at compensating for the suboptimal aim of the average human archer, even if it came at the cost of the rate of fire.

Even if that was not an issue, it would still not have put them at the same level as the Gru Bosch, the most elite of all the bowmen in the elven realms. They could hit a target from a hundred meters away while riding on a horse. The Gru Bosch were recruited from the sons of minor and sometimes upper houses, getting trained for over a decade to become proficient in the usage of their two bows. One was a recurved bow, best used when you had to come relatively close to an enemy and quickly needed to reload. The second was their longbow, which was almost as tall as a man, used mainly when you wanted to keep your distance from your opponent. Both these bows had been mastered, together with the use of many different arrows, from standard, to armor-piercing and even fire arrows though these saw rare usage, however, and very often they were not even brought into battle at all.

Her brother had joined the Gru Bosch and rambled on and on about how human crossbows were a disgrace to proper archery. She rarely listened to his speeches, but now he won't even speak to her at all after what Mynedrie had done to her.

She walked further through the armory, refusing to keep staring at the ranged weapons that seemed to mock her with her past. She instead returned to the sword section and took out two sabers, both identical to one another, with the hilt and its guard being made out of simple thin, dull, steel. The blades were sharp, with good curve and lightweight. The balance was in favor of the hilt. It weren't weapons for a noble-born lady like her, but it would do the job just fine.

Next up was clothing. With the place on her chest where it had been ripped apart by Drumir and hit by lightning afterwards, significantly burning away her clothes, to the point that she barely wore anything. Luckily, her new metal skin seemed to have prevented her from gaining any wounds because of it.

Lucky for her, there was an assortment of armor in the armory that was standardly used by the city's militiamen. Most of it was too big for her however, and so she would only wear an oversized piece of cotton armor to cover her exposed skin.

She would then move to the King and Queen's quarters, which was filled with exotic tapestries that covered the pale, chalk-covered walls and beautiful, carved furniture. There were also two big wooden doors, with each having a woman in a beautiful dress carved into them, with an equally stunning level of detail put into them.

Curiously, she moved to the two doors to see what they were hiding. As it turned out, it was a walk-in closet that was filled to the brim with the queen's clothing, shoes, boots, dresses for every occasion and even a few ceremonial uniforms for the sake of having or as a simple curiosity. Even before her family had disowned her she had not seen such a big collection and said to herself with a smile, “It’s gonna be good to be a queen.”

Chapter 6

With a shock, Mynedrie woke up, her last memory being that she got unconscious in the street, seeing the vague, dark silhouette of a man slowly approaching her, while fearing that she would in turn fall into the hands of Drumir. She looked at her surroundings; she was surprised to see that she was not in some kind of dark, unclean dungeon, but a comfortable bed in a nice, clean room. Still, her suspicions were not yet quelled. After all, things were not always what they seemed.

To the left of her bed there was a nightstand on which lay most of her belongings: her pouches with resources, the sheaths of her knives, everything was there, except her bow. She looked around the room quietly, mindful of the sound she was making, not wanting to disturb whoever had brought her here.

She looked beside the nightstand—nothing. Around and underneath the bed—nothing. Even in and around the closet and chair that stood right in front of it from where she was standing, yet there was nothing there, nor was there anything in the closet either, except for some fresh sheets.

Now being certain that Heartnailer was not in the same room as her, she knew that the next thing to do was to leave the room and search further in the rest of the building and the person that had captured her. Whoever had brought her here seemed hospitable to her, still she could not yet be certain of his motives.

So careful as she was, she sneaked out of the room and into a hallway. Right from the room's door was a staircase which led both upstairs and downstairs. On the left, at the end of the hallway, there was a window. From the right there was another room.

She considered carefully for a moment which way she should go. She chose to go downstairs, which seemed the more likely place where her bow would be, as upstairs was most likely the attic, where most of the old stuff was placed and forgotten about. Carefully she moved downwards, making sure that she would not be making any sound, nor any from the stairs.

Downstairs she ended up again in a long hallway. Behind her was the entrance. She softly tested it to see if it was open. It was locked. She turned around and half a meter from her left there was a closed door, with a lock on it. A meter or two in front of her there was another closed door, but this one was more inviting in its design, with white colored wood working as a frame to a crystal patterned window. There was no lock on it as well. Behind its glass she could see another door, through which the light that shone in the back garden shone through the door window, which distorted the view of what lay behind it. Yet the light shone perfectly clear through, lighting up the white colored hallway with a warming, late afternoon sunlight.

She walked towards it, minding her steps for any sound that they, or the wooden floor, may be making. At the door she put her hand on the handle, slowly pulling it downwards, letting it not make any more sound than it had to. From its small opening she peeked inside, seeing that it was a kitchen, which meant that there were knives nearby, which in the case of a fight could help her stay safe.

She opened the door just enough to get through it. She took a large cutting knife from the knife block on the kitchen counter. Now being armed with a chef's knife she continued her silent search for her bow.

Just left from the door she had snuck through was an opening to a room which seemed to be a long, white sheet covered table, with luxurious wooden chairs, with red cloth cushions nailed into them. Walking into the room she saw that the table was nearly twice as long as her. Yet with all the chairs that were standing around it, the room still had enough room left to fit stuff in.

She looked to her right and through the windows of the slide-in door that separated the table's room with the other one, she saw a man sitting on a comfortable looking leather chair, reading a book that had captured his attention with such intensity; she was not noticed by him despite being mere meters away from him.

She decided to slowly turn back to the kitchen, and from there turn back to the hallway and enter the room where he was reading, through the door she had skipped in favor of the kitchen's door. Slowly she opened it, knowing that any sudden creak or squeak of the hinges could give her away and alert him.

She slowly and quietly marched towards him, with the knife in her hand. Slowly she walked over the carpet of the room, which helped her in keeping her steps quiet, edging ever closer to her mysterious captor.

She made it to his chair, standing right behind it, hearing it make a stretching sound as he moved in the chair to get in a comfortable position. The mysterious man was deeply wandering between the sentences of the pages of his book, before something cold and sharp touched his throat and a hand wrapping around his forehead kept it in place.

“Do not move!” were the words that came from behind him. It was a woman’s voice with a fascinating accent, but he could not place from what language it originated.

“Wow, calm down. No sudden moves now,” he said with a calm voice, knowing that causing any stress now could worsen the situation severely for him.

“My bow. Where is it!?” she asked, while checking if he had any weapons on him.

As she searched him he answered, “Listen, I know that keeping a knife close to my throat, demanding to know where your bow is seems like a good idea...”

“I asked you about my bow, not life advice!” she interrupted.

“Then what would stop you from slitting it now?” he countered.

Mynedrie became confused and responded, “Cause if you do not tell me, I will!”

"Yeah, but then you still do not know where it is, meaning you would have to start to look for it yourself. If it is still here, of course," the man argued back.

For a moment nothing was said or done until Mynedrie retreated the knife from the man's neck and moved to stand in front of him. From the front she had a good look at the man, who had a young and bright face with thick, brown hair on top of his head and with dark-blue eyes looking back at her. His physique was hidden beneath his formal attire, but she could tell that there was at least some muscle underneath. From his expensive looking clothes she could surmise that he worked at the king's court, though what position was still yet a mystery to her.

He stood up and stretched his hand out to greet her, "Before I bring you to your bow, let me properly introduce myself. My name is Arlen, nice to meet you," he said, with a warm, gentle smile on his face.

Mynedrie looked at his stretched out hand, noticing that it

was his right hand. It was a known fact to her that humans preferred to greet with their right hand, as they tended to be right handed, whereas elves had a tendency to be left handed and so greeted mostly with the left hand. She found it odd to be greeted with a handshake from the right hand, even hesitating to shake it at all. Still, she almost immediately realized that it was about the gesture and not the hand itself. She let his hand meet hers, all while saying, "Mynedrie Kianistra."

"Mynedrie, beautiful name, but quite a mouthful. Is it okay if I call you Myna?" Arlen asked.

"Yeah sure," Mynedrie said, not really caring what he called her, as long as she got back her bow.

"Now let's get that bow for you," Arlen said to her.

They walked to the door in the hallway that had a lock on it. Arlen took a keychain out of his pocket and started to search through the many different keys that were on it. Putting the key in the door's lock, it only took a turn to open the lock, and with it the door, revealing a staircase that was nearly twice as long as Myna.

“Watch your head, or you might hit the cellar’s ceiling when you are going down the staircase,” he warned Myna.

They slowly walked downwards into the ceiling where much was laying, hidden in the dark, only semi-organised room. There was a forge, on top of which was a funnel shaped metal object, with the broader end facing downwards to the forge and the thinner end facing upwards, until it made a turn to the wall, where it went into and ended up outside. Left of it was a gate, just broad enough to let a small carriage go through it. Right of the forge was a workbench, above which were hanging all kinds of different tools.

Right next to that lay her bow, on a crate, together with a book. Mynedrie walked over to get her bow, expecting it first to be sure there had not been tampered with it. During her inspection of her bow, she noticed the book that lay beside it. The book was bound with oak brown leather, with on the front cover engraved in big golden letters, "Magical Weapons and How They Work. By: Drius Driar. Translated by: Dave of Greater Brada," with the image of a golden bow placed underneath that.

She noticed that a sleeve of paper stuck out from the top of the book, that had been placed between two pages. As she looked at the piece of paper, she could not help but think, well at least he is civilized with his books, unlike some others.

She placed the bow over her shoulder, so she could pick up the book and take a look inside the pages, specifically the ones where the piece of paper was placed, and sure enough she found that he was at a chapter which explained about magical bows of all kinds: their shapes, sizes and abilities.

On one of the pages there stood a paragraph about her type of bow, which read:

Light bows

Light bows, as the name suggest, are magical bows that

mainly weaponize light. They are primarily made from a white wood from a tree called pale oak, a rare tree that does not natively grow in the wild. Rather it is slowly grown by an expert alchemist. The tree then has to grow for around twenty years before the wood can be used to make the bows. Once the bow's initial shape has been carved, it is submerged into a liquid mixture of water, quartz, gold, silver and iron, which when combined with the bow's wood will fill it with strength and energy, which makes it near indestructible.

If such a weapon like this got scavenged or captured by any undead creation or dark magical creature, it won't hurt it directly. Rather, it has been recorded multiple times that anything or one that is powered by dark energy will feel a frustration that becomes more and more unbearable as they keep the bow in his, hers, or its, hands or claws.

The string that is used on such a bow is usually made of flax. While it may not sound as special as the pale oak, the process turns a seemingly simple plant into a material worthy of such a well crafted and powerful weapon. First they let the flax grow naturally until it is fully grown, then they will water it with a mixture of water, copper and quartz. This mixture will turn the already white flax even paler, with shiny sparkles on top of the fibers. Once the bowstring has been made and connected to the bow itself, it, unlike regular bows, never needs to be removed nor is there need for it to be kept dry. A string could be attached to a bow and left alone for a century and no draw strength whatsoever would be lost...

"Something you find interesting in that book? If you want you can borrow it for a while," Arlen interrupted.

Mynedrie placed the book down and said, "Thank you for the offer, but I have already read it once."

"You did?" Arlen asked.

"Yes, back when I was just a young girl. I would spend quite the time reading underneath my favorite tree," she said, while seemingly feeling nostalgic for those times, "It is better in its original."

"But enough about me, why did you bring my bow and a book about magical bows to a workplace like this? Should you not have it placed inside of a study?" she asked.

"Yeah, normally I would do that, but I needed both here for a device that I am attempting to create. Well at least to inspire me in making some potential adjustments," he said before moving towards his forge, where underneath a piece of cloth lay a shiny yet strange looking device. It was small, about the size of an outstretched hand in length, "You see, I wanted to know if there was anything in that book or in the way your bow was working that may help me in improving this weapon's design," Arlen continued.

"What is it called?" Mynedrie asked, looking puzzled at the odd looking device.

"I am not yet sure on a name. I am doubting between the name revolver, or waiting if any other name turns to mind," Arlen reacted.

"What does it do?" she asked.

"Well, it shoots projectiles that fly through the air at incredible speed, until they hit their target. Whereby the impact, if aimed correctly, can even punch through a knight's armor!" Arlen explained.

Mynedrie looked at Arlen as if he was speaking madness, saying, "With that small a thing, impossible!"

"I thought so at first as well before I started making it, but it is the truth. Tomorrow morning I will show you, cause it is starting to get dark outside," Arlen said.

Mynedrie contemplated if she should stay. It was clear that this man was smart, but the thing that he claimed his device could do sounded impossible if you did not have a strong enough bow or arrows with special armor piercing tips.

"I can show it right after we have eaten breakfast, which reminds me, if you want I can make you some food, you must be starving after having been in that fight with that raging beast," Arlen said. And just as he asked it, Mynedrie's stomach started to rumble, as if it understood and acknowledged the question and it was trying to put Mynedrie in a position where she could not say no.

"Well, I do have a little appetite," Mynedrie answered.

"Great, then I will jump into the kitchen and be right back," Arlen said. As he walked towards the kitchen, he quickly turned around back to Mynedrie and continued talking, "I do need my knife back."

Mynedrie gave back the knife to Arlen, who said, "Thank you. Now let's go back upstairs. If you want, you can wait in the living room, borrow a book from the bookshelf in my study if you want."

Mynedrie did feel enticed to do that, but a little voice in her head told her that she could not yet fully trust this man entirely, "Actually, I would like to help you with cooking. If it is alright, I mean you helped me this much so I would like to do something in return," she said. If he says yes, then she can keep an eye out if he does not put something in her food. If he says no, she has a reason to be wary of him.

"Well, there is not much you can do really, but I guess you could cut the meat if you want, I know that you are good enough with a knife," Arlen said.

They both moved to the kitchen where they prepared to cook together. Myna stood with the kitchen knife in front of a wooden cutting board on which she was cutting liberally salted steak into tiny pieces for the soup that they were going to make. Next to her was another cutting board on which lay an onion, a few string beans and two potatoes, together with a vegetable knife.

While she was cutting the meat and vegetables, Arlen meanwhile was filling a small, dark iron pot with water and carried it to the fireplace in the living room. There he hooked it right above the fire.

Mynedrie saw him doing this without any effort, even though such an iron pot with water in it was not light at all. She would guess that together the combined weight must have been close to 25 kg and he was carrying it with just one hand, not being bothered by the weight at all.

Arlen turned back to the kitchen and Myna quickly returned back to the cutting board before he could see her keeping him in her sight.

"It should not be too long before the water will start to boil and we can make soup with it," Arlen said, as he walked back into the kitchen.

He took a peek at the vegetables she was cutting, "You are still at the midway point it seems," Arlen said.

Mynedrie looked at the vegetables she had not even cut one of them this whole time; she had to come up with an excuse quick or he may start suspecting something from her. "Yeah, I uhm... I just got lost in thought, you know," Mynedrie said, while cursing herself in her head for saying something so stupid.

"Yeah, I know what you mean, seeing how you defeated that sea monster and have been laying in that bed for a whole day you can get..."

"A whole day!?" Mynedrie interrupted.

"Yeah, and not even a calm one. There has been unrest all throughout the whole city after those weird looking guardsmen came walking through the castle's gates. I happened to see one while I was in the market to buy food. His face was almost unhuman, his skin had become a light-purple complexion, with darker purple vein-like lines emanating from the crystal on his forehead. One man tried to speak to the unnerving guy, but all that came from his mouth was gurgles and sounds that almost seemed like they were words. For the rest of the time that I was there, he just stood there motionless, expressionless, almost like nothing was going on inside himself. That is if there still is a himself inside of him."

Mynedrie had trouble believing what she was hearing. The part of the soldier with the crystal on his forehead was believable. That, she had seen herself when she was in the castle and similar things when she hunted Drumir and lesser dark mages in all her years that she has been a Mageguard. Still the way Arlen described to her what the guard was doing was odd. Drumir used crystals like that to get himself bodyguards while he was making his latest and greatest monstrosity, but they only on rare occasions left his side and then not long or far away, yet she was told that the guard did both of those things.

This could only mean two things in Mynedrie's mind. Drumir looked to take over the city to either rule it, or try one of his creations on the city's populous and infrastructure. The latter seemed more likely to be the case, for she was well aware of his work, having personally been there when one of his creations had been led loose on her village. Just as he had done with many other villages and with many different kinds of abhorrent things, yet never had he tried to conquer anything, leaving almost as fast as he had come there after he was done messing around.

As they were just done preparing the ingredients for the soup, they moved to the pot with water that had started to steam. Arlen turned around towards Mynedrie, who carried the cutting board with all the pieces of meat and vegetables on it. "The honor goes to you," he said, while waving his hands towards the pot.

"Well, thank you, sir," Mynedrie answered back, while smiling at him. She poured the meat and vegetables into the pot, then some quick stirring with a long, large metal spoon and then all they had to do was wait until it was done.

Both of them went on to take a seat on the brown, leather couch. As they walked to the couch Myna noticed something laying on the leather chair that Arlen had sat on before she had threatened him with a knife. It was a book, specifically a book about the tongues of the elves. She picked up the book and asked in her native language, "A book about elven languages? Makes me wonder how fluent you might be in Moarla?"

He started to act a slight bit nervous as he tried to speak back,

"Uhm... I uhh... I not speak good yet."

Mynedrie started to frown and continued to speak in Moarla to herself, "Why are humans always learning Aelvin and yet practically not a single one wants to learn Moarla?!"

"What was it that you just said?" Arlen asked.

"Nothing. Just some... cultural rambling," Mynedrie answered.

"Well if you want to talk about culture, I have a few books that are about just that. Two about elven culture," Arlen said.

"Actually, now that you mention it, I would like to read something about history of mankind if it is not any trouble?" Mynedrie said.

"Not at all, I can understand wanting to learn about a new land you just came to," Arlen said.

They walked upstairs until they were at the second level, which was also the basement, which was also a study with a desk, on which lay papers, a notebook and a feather in a small pot of ink. There were also several pencils on the desk.

Left from the desk were several bookshelves that together created their own corner in the study. All were filled with books that were organized in genre and type of book.

To the right of the desk is a slide door made mostly of glass, but with a wooden frame. If you opened it you would end up on the balcony, from which you had a view of the castle and the light tower. The rest of the view was mostly the surrounding houses and some trees planted along the street.

They walked into the corner with all of the bookshelves. In this reading corner there is a simple yet comfortable looking couch, smaller than the couch downstairs and made of a light-grey cloth rather than leather.

Arlen took a book out of one of the bookshelves. It was a book on the history of mankind and the cultures of its many people groups, 'The History of Mankind and Its Peoples. By: Bertold of Nordmark.' He said while reading the book's front cover, "Anything that you want to know specifically?" Arlen asked.

“The history of this town would be useful to know,” Mynedrie asked.

Arlen started to plough through the large book’s many pages until he reached the page that he needed.

“Here it is, 'The History of the Kingdom of Drun, from 297-500.' It says here that it was founded by Jacob the First, a noble in the time of the great Maktow Empire. Building the first settlements to monopolize trade around the mouth of the Drune river. As the town was slowly growing into a usual city, the empire it was a part of started to crumble and rot until finally collapsing in 384. Creating a power vacuum, which allowed our humble little city to become an independent city state with its own king. During the reign of its first king (who was called King Otto I of Drun) he started several building projects to protect his vulnerable city against any form of assault. Most notable are the massive walls surrounding the city, which replaced the relatively small walls that preceded them over time. They were finished in the year 392. After a nearly twenty-four year long reign, he was succeeded by his son Erik I, who continued further construction on the city throughout his reign mainly the basis of what would become the city's sewer system (which I improved on).

In the year 301 he partook in a war against the southern realm of Catinga, which had started a war against the island of Tharana, which was Drune's biggest trading partner in that region, for it was mainly their ships that passed through the mouth of the Drun river to do trade with the rich lands that were to be found along the river's banks.

The war ended badly for King Erik as Drun was encircled by the ships and armies of the Catinga, which were armed with massive reptilian creatures, which were slim and laying low to the ground, yet were also long, ten meters in fact, from head to the tip of its tail.

Slowly, the creatures moved to the wall. With the defenders shooting with their bows and crossbows towards the massive creations of nature, yet doing no damage, for their scales were too strong.

As for the people that rode the great beast, they were protected by a small and thin metal box with just enough room for two men to lay in. It only had two small holes through which the two men inside could look at.

As the creatures reached the walls of the city, they pushed themselves upwards until they stood on only two legs. When that did not prove enough, the creature pushed itself even higher with its broad tail reaching a height of eight and a half meters if its long snout looked upwards. Sadly for the soldiers on the walls it did not, for it used its powerful maw to shred any men unfortunate enough to be caught by it.

When the beast stood upright against the walls, the soldiers that had been following it started to climb onto the net that lay on top its scales, climbing onto the walls, to fight the already struggling soldiers on top."

Mynedrie stood in deep thought, contemplating the

information she just heard. As she did this, Arlen asked her, "Why did you want to know this city's history?"

"Because of Drumir," she responded while still attempting to contemplate.

"I do not follow?"

As she paced around the room she started to explain, "Drumir is a maniac, but he always chooses his targets with a reason. He had days to get out of his cell when we were still on the ship, yet he started his escape only when he was closest to this particular city. If I want to know why, I first got to know this place and understand its history." As she continued to try to piece together what Drune's possible usages could be she would be distracted by a flash of light coming from outside. At first, she thought it was just a flash of lightning until she realized that the color was off. This detail was also noticed by Arlen, who walked with her towards the balcony.

Even in the dark of the night, they noticed that a cloud was moving circular around the lighthouse, its location revealed by the torches that lighted up the courtyard hidden behind the castle's walls and the massive branch which burned like a torch right on top of the lighthouse's highest point before it started to move downwards to the balcony where its roots were planted.

Mynedrie turned to Arlen and said, "Whatever it is that he has done, he has already begun!"

Chapter 7

Mynedrie and Arlen walked through the busy streets of the city's mall. Merchants from faraway lands were there to sell their wares. You had merchants who brought something as spectacular as a rare beast or something as a specific spice that refused to grow in the local climate.

The interior of the mall went up three levels, all filled with stores that sold a wild array of different things, both simple and exotic. On the top of the mall was a glass dome through which light entered the building. A status symbol of the ruler it was, with the glass being framed around gold. Some of the rectangular glass plates could be opened to let out any heat that would build up inside the building and closed if it became too cold, like in the winter, during which they would pull a steel frame with a curtain attached on it entirely over the glass. Preventing snow and hail from falling on the glass and keeping the heat inside.

Mynedrie was enamored by the many stalls and shops that

were inside. She turned towards Arlen and said, "Arlen, this place. It's incredible!"

"That it is indeed," Arlen responded.

They walked across a corrupted militia member who seemed to be staring simply forward, not really focusing on anything in particular. Still, the presence of one of Drumir's unwilling lackeys was enough to send a chill through her body.

"Are you certain this cloak I'm wearing will keep me hidden?" she asked.

"Of course, without seeing your ears they just see a pretty lady in a cloak. As far as they are concerned, no elven ears no elf," Arlen responded.

Mynedrie pulled her hood further forward to be certain they were not seen. Elves used to be not uncommon in this city as trade was going well between Drune and the Elven realms until a few years ago when a new emperor took control and gradually decreased trade with the human cities.

"Look, here it is," Arlen said.

They walked into the store which was filled with spears, arrows, maces, halberds, axes and of course swords and knives which had their own racks in the back of the store.

While Mynedrie examined the wall and racks with all the knives that were behind a counter, while the hilts for such knives were placed neatly in a rack just beneath the counter, she asked one of the store employees to closely examine the knives that seemed interesting to her.

Meanwhile, Arlen was looking around the store himself, examining whetstones, that lay in a wooden rack that was placed in the middle of the room. His attention would be caught, however, by a box made out of wood with the shape of curved knives, with luxurious looking handles, both with their tips pointing to the right while tilted slightly upwards. He walked closer to the box, which had a big iron lock on it. In the middle of which was a big keyhole. The upper side of the box was carved as well. It showed a hooded figure fighting against a massive beast thrice his size, with nothing but a knife in each hand. Stabbing with the one in his right hand the creature in the neck, while the other arm was stretched out, making the figure seem to stand proud.

While the craftsmanship of the box alone was enough to make him consider buying it, he had to remind himself that he was here for the knives. He saw that a bell was hanging right above on his left, under which stood the sentence, 'Ring if you need someone,' with an arrow above the sentence aiming towards the bell. Arlen then rang it. Soon afterwards a broad, muscular, man with a full grey beard stepped out from behind a curtain. The man was half a head bigger than him and Mynedrie and wore a leather apron which was covered with black and grey smut.

He came to the glass counter where Arlen was standing and asked in a harsh but kind voice, "Hello sir, what can I do for you?"

"Yes, hello, I was wondering about this wooden box that you have on display here?" Arlen asked.

"Sure, what about it?"

"Well, for starters, why are they hidden in this box? Seems to me not like the best way to get someone to buy a knife without seeing the knife first."

"Oh that, that's because, about a week ago someone tried to break the glass and run off with them. The poor fool did not realize how sharp glass actually can be and cut himself in the attempt. It's a good thing that he survived because otherwise I would have had to pay for the replacement of the glass and not him. You know that reminds me of when another..." As the smith continued to talk, it would eventually be noticed by Myna, who noticed that Arlen was stuck in a conversation with the man, seemingly unable to get him to stop, "Oh no," the salesman behind the counter said.

Myna turned and asked, "What's wrong?"

"It looks like my brother is telling one of his never-ending stories again. That guy must have asked him something and when my brother starts speaking he does not stop," he explained. "Hey Gregor, how about you tell him about what he wants instead of your life's story," he continued.

Gregor looked angrily towards his brother, before grumpily saying, "What else did you want to know?"

"If you can show them to me," he asked.

"Hey Timothy!" he shouted to his brother, "I need the key for the wooden box!" he continued.

Timothy, the slightly thinner, though just as tall and muscular of the two, said to Myna, "Excuse me ma'am." His full black beard dancing on his face as he spoke. He then walked towards his brother Gregor, while taking out the key from underneath his brownish shirt, which hung around his neck like a necklace.

He placed the well-polished key, which was crafted with great attention to detail, the key's head being in the shape of the two knives points meeting leaving a hole in between the knives and the lower end of the bow.

He placed the key inside the lock. As he turned the key, you could hear the inner mechanism make the sound of metal clashing and rubbing over itself before finally releasing the box so it could be opened.

Timothy opened the box, revealing the beautiful weapons inside. The blades themselves are about the size of Arlen's outstretched hand. The spine of the knife being filled with deep serrations, up till around a third of the blade's full length, where it curved towards its point with a fake edge. From there the edge of the knives continued until they reached the choil of the blades. The ricasso was thin, only half a centimetre long before reaching the quillon. The handle's wood mimicked the metal's pattern of curving lines, running all over the surface, only these were shades of brown, not grey. It is a smooth handle along which were three rivets made of the same metal as the blade itself. Last was the pommel, which was in the shape of a slight curvature inside of the handle itself.

"Wow," was all Mynedrie could get out of her mouth as she

gazed on to the exotic knives.

“How much are they?” Arlen asked looking just as amazed.

“25,000 coins,” Timothy said.

“25,000!” Arlen said in disbelief.

“Is that much?” Myna asked.

“Much, that is enough to give a group of forty men a good knife each,” Arlen explained. He turned towards the two brothers and said, “That’s an outrageous price!”

“Outrageous! Do you not see the materials it is made of!? The serrations on its back!?” Timothy started explaining.

“So what about those serrations, they just seem like some kind of useless decoration,” Arlen argued.

“Not true, they serve a great function, for you see they work kind of like as the teeth of a saw, very useful when you are in the wild.”

“That still does not justify such a price for two knives,” Arlen argued.

“Well, it is not just the knives, it is also the belt with the sheaths for them as well," Timothy said before taking out the upper part of the box in which the knives lay, revealing a separate area in which the belt and sheaths were in. Made from a dark-brown leather, it showed to have been made with absolute craftsmanship.

"You of course also get the box," Gregor added, standing with his arms crossed behind the counter.

"They seem like they're enchanted?" Myna asked.

"We thought so as well, but the guy who gave them said they were just forged with advanced smithing and high quality materials," Gregor explained.

Arlen stood with his mouth rubbing his chin, contemplating, before saying, "10,000."

Both brothers looked insulted at the offer, with the black bearded Timothy saying, "10,000? That is not even half the price!?"

"Yet still you walk away with a full profit, seeing as it was 'given' to you. Which brings up another question: why would someone give knives of such quality just away?" Arlen argued.

"Well... he did give it to us technically," Timothy struggled to explain.

"Technically?" Arlen asked.

Timothy and Gregor looked at each other like they had just realized something. "There is something wrong with these knives, is it?" Arlen asked.

"Of course not! We would never sell anything to the customer if we did not think it was of the highest standard; it is just... we played a few card games with a merchant who was docked in the harbor. He was going pretty strong for the most part, beating us mostly, until he had put in this box and then we won the match easy... too easy," Timothy said.

"So you ask 25,000 for a box with two mysterious knives that you won 'too easy.' I'm starting to think that 10,000 is even too much," Arlen argued.

Timothy looked up, realizing that in his small moment of

introspection he made the mistake of speaking aloud what he thought. This had not gone unnoticed by his brother, who bumped his shoulder with his fist, while saying, "Nice job idiot, do you want us to lose money!"

He looked at his brother sternly who did the same back to him. It went on and on, until Myna interrupted by speaking, "Well now gentlemen, there is no need to get angry about it. I'm sure two strong looking men can understand the need for a woman to protect herself," she walked closer to Gregor, who still stood on the other side of the counter, which she bent over, while she continued speaking, "Especially you big man."

"Who me?" he said while laughing insecurely.

"Of course you dummy. You look like you could protect me, with those strong muscles of yours," Mynedrie said.

"Well, they are big," Gregor said with increased confidence.

Her speaking decreased to a passionate whisper, while both Arlen and Timothy looked on, "You know I often feel scared when I'm on the street 'cause I don't have someone as strong as you to look out for me. With those knives, however, I might feel just a little bit safer. You want me to feel safe, right?"

He looked at her with dreamy eyes, into hers before turning his towards his brother and saying, "I think she deserves a discount."

He then turned his head back towards Mynedrie, who said, "Thank you, love," she then retracted back to her side of the counter, while also blowing a kiss towards him, who in turn acted like he had caught it in the air.

Timothy looked with anger and disbelief at his brother before turning his head towards Arlen and said, "Alright, you heard the lovebird, you can have the knives for only 10,000!"

Arlen held out his hand to shake on it, with Timothy meeting it with his hand. As they shook their hands firmly, Timothy said in a grumbling tone and sour face, "Well played, both of you."

As they walked out of the store with the sound of the two brothers arguing loudly about the knife's decreased price, Arlen asked Mynedrie a question, "Soo... what was that you just did back there?"

"Oh that, just a little trick my older sister taught me. Afraid I'm not as good at it as her. She could have gotten those knives for free, with almost no effort," Mynedrie explained.

"Well for what it's worth, I think your flirting was perfect."

Mynedrie laughed warmly towards Arlen while saying, "Aaaahhhh... thank you."

For just a few seconds, a short moment of awkward silence seemed to have grown between them. A silence that Arlen broke by answering another question, "Do you have any other brothers and sisters or is it just the one you mentioned?"

"No, I have a total of seven siblings: first is Aisyane who is a really sweet and nice woman, very caring; next is Yaismane, who is the flirt I told you about; then we have my brother Daimir, quite the adventurous type; then we have me, followed by my sister, Diania, who just like Daimir is adventurous, but with added curiosity and bratyness; last but not least we have the twins, with my brother Muinak and his younger sister Muinin. Muinak is a great companion to everyone that comes his way, Muinin however is more shy, preferring much more the company of books than people," Mynedrie explained.

They made their way through the bustling marketplace that was just outside the mall. Where you could get mostly special and/or luxurious items in the mall, you could get more common things in the marketplace: vegetables, simple clothing, little trinkets—that kind of thing. Most of the people that have a stand here came mostly from the northern slums, even if the clothing did not make it look like that, with most, if not all, wishing that one day they can get their own little corner in the royal mall, but for now they were employed to the merchants that were providing their stalls with wares for them.

"We still need to find some armor to replace the one I have left inside of the castle," Mynedrie said to Arlen.

"I know, but we simply do not have any armorers that have any armor for women in stock, elf or human," Arlen explained.

"Then we will need to find an alternative solution, 'cause without armor I will be far more vulnerable than I like to be. If only we knew someone who has the skill to make any?" Mynedrie contemplated out loud.

"Well, I don't have much experience in making armors, refitting however, that I can do for you," Arlen explained.

"Sounds like a good plan to me," Mynedrie said.

As they walked through the market checking the little stands for anything interesting, Mynedrie noticed the giant dome-like building, which she soon realised could give her a good look towards the city's castle. She turned towards Arlen and asked, "Hey Arlen, what is that building right there?"

"That, that is the Divine Tree Temple of Drune, the religious centre of the city, why do you ask?" Arlen said.

"I want to take a closer look," she answered.

And so they set out towards the temple. It was large with it being a large square building that went up ten meters, on top of which was a dome, which was nearly as tall as the building itself. Making it the second largest building in the city only beaten by the lighthouse, which was five times bigger than the temple. A fact that the more religious of the city would have unseemly if it was not that the light tower had been built centuries back when the city was still just a small town.

Mynedrie looked around the building like she was searching for something; this was noticed by Arlen who asked, "What's wrong?"

"Nothing, just trying to find a way towards the top of the building," she explained.

"Well, besides the fact that that sounds quite disrespectful, I'm not sure you get away with climbing a twenty-meter-high building in the middle of the day," Arlen argued.

"Perhaps if we try through the inside?" Mynedrie asked.

"Perhaps," Arlen said.

They walked into the temple. Mynedrie was enamored by the absolute size of the interior, which was almost one gargantuan room, with walls filled with masterful works of art, pillars made of black stone, with white-colored cracks in them. On neither of these constructions, however, were any kind of torches to give light to the great room. Instead, all of the light inside of the temple came from one source: a pale, stone carved, bowl filled with a shining liquid, laying right in the middle of the room.

Mynedrie looked with her eyes wide open at all the beauty of the interior of the building, overwhelmed by the sheer size of this hoard of art and wealth stored in one place so central to the city.

Her examining of the building did not go unnoticed however, as they were approached by an old acquaintance of Arlen. Her name is Sophia and she is the devout and zealous teacher of the temple, giving its members spiritual and life advice, as well as taking care of the temple, the liquid in the bowl and of course the many traditions of the temple and its visitors.

Sophia almost immediately recognised Arlen and for the hooded lady beside him that examined the great inner sanctum of the temple. She walked over towards the two. Wearing the white colored robes of the temple, while standing proudly and upright she asked with a smile, "Arlen! What a pleasure to see you here once more. I'm afraid, however, that our morning rituals already have ended."

"Sophia, no surprise finding you here," Arlen said, his discomfort being noticeable in the pale blue light of the temple, while desperately hiding it behind a fake smile to keep up a good and proper appearance.

Sophia looked with a glaze at Arlen that seemed to go right through him, searching for something, what Mynedrie could not answer however. One thing was clear however: there was a history between the two, one she was not sure of if she had the right to pry into.

Sophia turned her head towards Mynedrie, asking, "And who might you be, miss?"

Mynedrie struggled to come up with an answer. She knew however that she had to come up with something. Before she could speak however, Arlen chimed in at the last moment saying, "Her name is Myna. A guest of mine who wanted to see the temple and some of my inventions and improvements of the city, if we have the time."

"Interesting, that a simple woman from out of town would come here for your little projects that you connive in your attic," Sophia argued.

"Well, she came here on the behest of her liege," Arlen answered back.

"I see," Sophia said while judging Mynedrie with her gaze. "Well Miss Myna, I do not know how things work wherever you come from, but here we take our hoods off while we enter the temple!" she continued.

A rush of angst came over Myna as she realised that the second she took off her hood people would notice her elven ears, with all consequences that entailed for her and Arlen. Arlen again however was the one that saved her disguise at the last moment, saying, "I'm afraid that she would not be able to do that. You see, dear Myna here was born a little different than most others, if you get what I mean and she is quite self-conscious about it at the moment, so if you could let it slide this once, please."

Sophia contemplated their request for a moment, before she responded, "I guess it does not go against any scripture, directly, and seeing how it is her first time here I will give her an exception. This one time, understood!" She pointed her finger towards Mynedrie like an upset mother or disciplining teacher. Myna responded with, "Yes, understood," trying as hard as she could to hide her accent from her.

She retracted her hand and said, "Good, now I hope you will find this place as inspiring as I do." She then left the two to go back to her duties. As soon as she had left they started to search a way to the top. There was a way to the temple's dome that Arlen knew of. Problem was that the entryway to there was locked. With no key to get inside they had to get creative.

Before they had found a solution to their problem Mynedrie asked Arlen a question, "Does it not feel wrong to lie in your place of worship?"

"I did not lie," Arlen responded, "I merely left out the specifics to her."

"But you said my name was Myna?" she asked confused.

"Yes, because that is the nickname I gave you," Arlen responded.

"But you said I was your guest? I may not understand human cultures well, but I know you don't take them out shopping for knives," she asked again.

"Perhaps... a bad chosen synonym, but not even close to a lie," he answered again.

Mynedrie did not know how to feel about this. She was neither a good liar, nor a fan of lying, but the way he worked around it with simple technicalities was, even if you disagreed with its usage, impressive. If a bit concerning when it came to someone she started to deem a friend.

They continued their search for a way to the top of the dome. Standing in front of a slab of stone, on which was carved the depiction of a mighty battle raging between two armies. On the left side stood a mighty and noble looking general, who seemed to have in his hand a double-edged sword, which he pointed towards the right. There on its right side was carved a frightened, obese, looking general, who was borderline ready to retreat and run away from the battlefield.

"Found any way upstairs yet?" Mynedrie asked.

"No, not yet," Arlen answered.

"This is getting frustrating, there must be a way to get there?"

Arlen looked around thinking for a solution. He then noticed the space between the mirrors and the walls. He turned towards Mynedrie saying, "You could try climbing the walls."

"Climbing the walls?" she asked, confused.

"You see that space between the mirrors and the wall? If you go through that you get in the area of the temple that is called the orb. From there you will find it relatively easy to get onto the roof, as there is a hatch that you can open, from there it is a simple climb up the dome and you will have the view that you are looking for," Arlen explained.

"Well, I could make that climb, but I doubt I could do it without people noticing," Mynedrie responded.

"So we need a distraction," Arlen said.

"That we do," Mynedrie said. "Tell me what is going on in this stone carving exactly?" she asked.

"Oh, it depicts the battle of Drukbrik. A small village at the time, its lord heard that a terrible army was approaching. It was called the army of Red Morning, for everywhere they went they would leave nothing but the burned buildings and bleed out bodies of their victims. Led by a rich, powerful, corrupt and cruel count who, after being cornered in a high stakes political game, was on the verge of losing his noble position. In a desperate attempt to keep his influence he would hire mercenaries in secret, to get rid of his biggest rival. He attacked the man's castle, pretending to be the men of a minor noble, loyal to the rival, to lower their guard. The plan was a success. However, the secret had been exposed. Now that everyone would soon know him to be in rebellion against the crown, he would use all his wealth to hire more mercenaries and would free anyone in prisons of the places that he attacked, as long as they were willing to fight for him that is. His force had grown beyond such a size and plundered so much wealth that he had become richer than he was ever before; he seemed absolutely unstoppable.

Until Baron von Dubrik came and fought the powerful and cruel count, using an army that consisted of the remains of other armies that had been sent to squash the rebelling count. It was not looking good for the good baron however, as he was outnumbered three to one and most of his soldiers were inexperienced. This did not stop him however, as he knew that help had been prepared in the skies above. For just as the battle began, rain was starting. This was no ordinary rain however, as this rain consisted of the divine liquid, that also happens to lay in that bowl; it strengthened the defenders' will, doing nothing to the Red Morning troops, giving von Dubrik's army the strength and courage it needed to defeat their enemy!"

"Fascinating," Mynedrie said in a hushed tone.

"Why did you want to know?" Arlen asked.

"Now that I know who the bad guy is, I know against whom I can put my foot to lift myself up, so I can grab its upper edge; from there I can grab that white cloth hanging there and make my way up to the top," Mynedrie explained.

Arlen looked worried at the white cloth that the temple uses to make shadow shows about the many accounts that could be found in the tome of trust. He then looked worried at Mynedrie and said, "I'm not sure that it will hold."

"Well, it is either this or we can kindly ask Sophia if she lets us go to the top," she said.

Arlen quickly glanced towards Sophia before saying, "You would have a better chance flying."

"Still, we will need to distract the public with something if we want to remain undetected," Arlen continued.

Arlen and Mynedrie looked around, contemplating and planning on how to distract a crowd of this size. The morning ritual was over and the midday ritual was still far from starting, yet there were more than twenty or so people walking inside the temple; some sitting on the benches, others were examining the art that was to be found in the temple, there were also the priests and priestesses that were walking around. Some were talking to the visitors, others were having discussions with each other.

It was clear that a major distraction would be needed to get them out of the temple or at least not notice Mynedrie. Luckily Arlen had just the thing: he took out a sack and grabbed something out of it holding it firmly in his hand before walking towards the entrance and said, "Everyone, there is free money outside!" He then threw his handful of coins outside for all to grab.

Many people rushed outside to grab whatever they could, but another group moved towards Arlen to scold him for his action. Arlen at that moment defended himself verbally holding the outraged group's attention as long as possible.

With the public distracted, Mynedrie was free to climb. She jumped on top of the stone carving towards the white cloth using it as a rope to climb upwards towards the edge that was in between the sphere and the wall. She climbed as fast as she could; however, from the corner of her eye she could see a young girl looking at her with bewildered eyes. The two stared at each other silently, only the commotion from outside filling the temple with sound.

Mynedrie put her index finger in front of her mouth, all while pushing softly air through it. The girl shook her head and waved her right hand, signing that she wanted something from Mynedrie.

Mynedrie looked around, looking for anything to give. She then realised that she still had a sack filled with coins in it. She carefully took out one of the coins and showed it to the girl, who was waiting for a better offer. She took out another coin holding out both of them. The girl still stood unflinching. Realising that she was running out of time, Mynedrie relented and threw the whole bag towards her. The girl weighed the sack in her hands, listening to the tingling of the coins inside, before signing her approval to Mynedrie and making her leave from the opposite side of the entrance from where the commotion was happening.

"Well played kid," she whispered to herself.

She climbed further until she reached the edge. There she

pulled herself upwards until she was safely behind the rails of the walkway around the sphere. She checked if no one was around, before she made her move towards the hatch. She opened it and was greeted by the rays of the shining sun, surrounded by blue skies and the occasional small cloud.

She noticed that she no longer heard any commotion from downstairs, perhaps she did not hear it from such a height on top of the roof, or perhaps it had ended. Either way, knowing what was going on was in her best interest. She carefully crawled to and looked over the edge, seeing that the group that had gathered in front of the temple had dispersed with many continuing their business inside of the temple.

Mynedrie knew she could not get back downstairs through the same way she got up. That was a problem she could solve later however. First she had to go and look at the castle and its defenses.

She climbed up the temple's dome, reaching the top on which was a small, thin spire made of metal, from which rode a cable down the dome onto the roof, where it lay neatly rolled up. She wondered where it could possibly be for before returning her attention to the castle.

It was on the top of the temple she could truly understand how big it was in comparison to the castle; only the lighthouse next to the castle was taller than either of them.

She made herself as small as possible and kept herself partially hidden behind the spire to make sure she was not noticed by either a guard or a civilian who would immediately call the attention of others if they happened to look up.

From the height of the temple's dome she could almost directly see into the outer court, where she noticed the roof of a tent that was not there at first. Second, she noticed that the damage and debris of the battle against the corrupted Drillin still lay all around. While there was damage, the key defenses of the castle did still seem to be fully operational. This would make any infiltration or attack difficult.

Mynedrie was about to go back down to the roof of the temple when she suddenly saw a large number of people run away in fear from the city's harbor that lay next to the castle. She tried to take a better look at the harbor; her view was however blocked by its dark wooden buildings and ships that had been dry docked for repair. From what she could gather from the small bits and pieces that she saw it was clear that it was being raided not by a foreign force, but by the militiamen of the city, most likely those corrupted by the crystals of Drumir; what they were after exactly was still a mystery to her.

Chapter 8

Mynedrie rushed to Arlen who waited for her at the steps of the temple's entrance. "Arlen, something is happening at the harbor!" she shouted towards him while sprinting with leaves sticking all over her cloak.

"What do you mean? Who could possibly attack the harbor? The great chain has been raised. And why are you covered in leaves?" Arlen asked.

She quickly glanced at her cloak before saying, "Uhh... not important right now. Come, we need to go there!" She grabbed his hand to pull him with her to where they needed to go, "We will need to quickly make a stop at your home to get my bow."

As they ran towards Arlen's home they would see people running in the opposite direction of theirs. One man even grabbed Arlen, stopping the duo for a moment and franticly warned them, "Don't go there. That grey woman and her soldiers are taking everything in sight and attacking anyone in their way!"

"A grey woman?" Arlen tried to ask the man before he continued running away.

"Must be another one of Drumir's monstrosities," Mynedrie said with contempt before continuing on their path.

As they reached Arlen's house they were able to hear that the plundering of the harbor and the screaming of the victims could be heard clearly. The screams of men, women and children was only silenced when Arlen closed the front door. Mynedrie quickly sprinted up the stairs to get her bow; Arlen meanwhile laid down the box with knives beside the door, then he ran to the door of his basement. He quickly searched through his keychain to find the one that would open his basement. As he found the key and was about to open the door Mynedrie came storming down. When she saw that Arlen wanted to go to the basement she asked him, "What are you doing?"

"I got something that might be useful in this fight. In the

meantime you can put your new knives to work," Arlen said before going downstairs.

Arlen rushed to the closet where he had put it in. When he opened the closet, however, it was revealed that the weapon he had made was disassembled. Looking at the disassembled pieces he remembered that he had taken it apart to repair it after a failed experiment had damaged it. "Arlen, you fool!" he whispered under his breath to himself.

"Arlen, you ready?" Mynedrie asked.

"I'm coming," he shouted back. He then ran up the staircase ending up back in the hallway of the house.

"Where is the weapon?" she asked.

"I'm afraid it is still out of commission," he responded.

"Well then, we will just have to test out these two pretty things, won't we," she responded while finishing putting on the knife belt. When she had put it around her hips she placed the knives inside their sheaths and continued speaking, "Come, we have a city to save."

As they opened the door the screams and cries of people and the plundering of property could be heard by them. It was nearly ear-deafening, with the sounds only getting more intense as they came closer and closer to the harbor.

There they would see the true extent of the carnage. Carts and crates had been knocked over; the stuff inside being taken by the plundering militia members. Blood painted the walls and brick road. The spatters, slashes and pools revealed that there was little to no self-control from the militia that came in and attacked the sailors and merchants that were going about their business before the plundering started.

In the middle of the little marketplace, which seemed to have been built in front of the docks, there was an old well, with right next to it laying the pile of plundered goods that the corrupted soldiers had taken. It was not very varied however, as most of it seemed to be iron, steel and iron ore that had been mined in the mountains in the east, which were overlooking Drune, instead of the more valuable and exotic cargo that they could get their hands on.

The mountain of iron was protected by two guards until they had been shot by Mynedrie, who together with Arlen then walked further in the harbor. This revealed that in a street on the right side of the little, grey, stone-covered plaza, there was a little twelve year old boy covering his eight year old sister, protecting her from a hit that was about to be struck by a corrupted militiaman, which would have slashed open the boy had it not been for Mynedrie, who came in just at the last minute to shoot the attacker before the hit could be struck. The boy opened his eyes after hearing the soldier drop to the ground. He looked behind him seeing the lifeless body of the man who wanted to attack him, a man from which power over his own actions had been taken away by the crystal on his forehead. The boy then looked around, seeing her and Arlen, with Mynedrie standing with her bow in her right hand. The boy told his sister to get up and move. Before leaving, he quickly told Mynedrie, "Thank you, ma'am."

Mynedrie smiled and said, "Not a problem child, now go get out of here, it is not safe."

The boy and his sister ran away from the little marketplace into the city itself, while Arlen and Mynedrie walked towards the docks where most of the plundering was taking place. Mynedrie would along the way shoot two other guards that were walking upwards to the big pile of metal in the plaza, carrying as much looted metal as they could.

From on top of the small slope they had a clear sight on the situation, up to twenty maybe more soldiers were plundering the harbor for everything it was worth. Uncontested they could go to work as there were no more people to pose any challenge to their whim, alive anyway. They were indeed led by a grey woman who ordered them around and what to target.

Mynedrie prepared a shot to be fired from her bow, carefully aiming so as to hit her target perfectly.

"Are you sure attacking so many enemies at once is smart?"

Arlen asked.

"Do not worry, first time I fought one of Drumir's creations, I found out that usually aiming for the biggest takes down the whole army," Mynedrie explained.

"What, army? What do you mean?" Before he could finish Mynedrie shot her radiating light arrow towards the commanding grey woman. It was certain that the arrow would have hit its mark if it had not moved at the last minute, dodging it with a swift and smooth turnaround.

"Hello cousin!" said the smiling woman.

Mynedrie's eyes widened in disbelief, saying, "Stria?!"

"I'm surprised you recognized me. I figured your little brain could comprehend only the simple things like how to please your father," Stria said, with a smug sense of superiority.

"I'm just surprised that you are still alive after Drumir tricked you into freeing him. Figured he used your corpse in one of his experiments; it seems I was not far off in that regard," Mynedrie responded.

"What, this little makeover. Just his way of thanking me, for freeing him from that metal box he had found himself in. I admit that the process was painful, but a girl gotta suffer to look her best, though I suppose you would not know much about that in that regard," Stria said.

Arlen meanwhile, looked curious at one of the militiamen's bodies. It was by no means pleasant to look at the corpse of a man you did not know, slain for actions that were not in his control. Still it was in Arlen's nature to notice something on the militiamen's armor. That being the royal crest that was in the shape of an anchor had been smeared over by black paint. The smudge that had besmirched over the crest was a circle, in which was a form that Arlen could only guess had to represent some kind of a bird. He quickly glanced at all the other militiamen that were taking position beside Stria, preparing for the battle that was to come. Each had the symbol smeared over their uniform, slightly different on each, like the men that wore them had been told to smear it on themselves with not even a mirror to help them, but the same symbol nonetheless.

"The paint job on their uniforms, did you do that?" Arlen asked Stria who was wearing the symbol as well, but applied with far more attention to detail and style.

Stria looked surprised at Arlen as she did not expect any humans to stay around, let alone ask her about her family's crest, "Excuse me, who might you be exactly?" Stria asked.

"My name is Arlen, an adviser to His Majesty," he said.

"Yes..." Stria said, while waving her hand in a circular motion, expecting a last name from someone claiming to be an advisor to the king.

Arlen knew what she was looking for and he answered her question while holding back his discomfort as much as possible, "I have no last name, yet."

"A commoner with ambition, well I am surprised. I suppose you are here with my cousin, like some puppy following their owner. I admit Mynedrie, I assumed that you would never find any man willing to put up with you."

Mynedrie felt her words sting her heart; she did not want her cousin to see that however and said, "Just answer his question."

Stria could see through her niece's act. A little more pushing and she might just break. "The mark on their suits is supposed to represent my family's crest. Though it seems that letting them put it on one another was not the best idea. It seems that the crystals on their heads kill any creativity they have, if they had any to begin with." Stria made direct eye contact with Mynedrie and continued, "Just like how they killed your daddy!"

The second she said that Mynedrie could feel something snapping in her mind, before she knew it she shot an arrow towards her aiming for her niece's head specifically the stone buried in it. Such stones she had seen before during her Mageguard career. The arrow was quick but Stria was quicker, dodging one arrow and blocking another with one of her sabers. The third one however was an explosive light arrow. The extra few seconds that it took her to charge it was enough for Stria to shout an order before jumping towards cover, “NOW!”

Almost out of nowhere several crossbow armed soldiers revealed themselves. They stood on top of the roofs of the harbor’s wooden cranes and buildings. The corrupted militiamen may have become less intelligent and skilled in favor of increased strength and no longer bothered by such things like morale, yet not doing anything for hours at a time, seemed to have been a skill that improved after the crystal had attached itself to their forehead.

The guards opened fire on Mynedrie, who had just released her own shot. The shockwave took down some of the militiamen permanently and pushed several others to the ground. Mynedrie then ducked to the ground before retreating back to the water well. Arlen followed her lead. The entryway however had been blocked by a wall of ten to twenty militiamen with the amateurishly painted crest on their uniform. They blocked the way with the points of their spears and halberds pointed towards them.

They turned to the left where they would be saved from the hailing bolts that were being shot their way. Arlen turned to Mynedrie and smilingly said, "Okay, that was fun. Now what?"

Mynedrie looked just around the corner to see what they were dealing with exactly. She counted four men, each on a separate roof armed with a crossbow. She also noticed that around five men were coming towards her. Some had small wounds that they had gotten as a result of her explosive arrow throwing debris around the air. She knew that more soldiers were still unharmed after the explosion and it was unlikely they were still in the harbor, meaning that they were approaching her from the other side, readying to surround her and Arlen.

She retreated back her face from the corner and started

thinking about any way out. She could try to climb up the wall but she would turn herself then into a sitting duck, while Arlen would be standing alone on the street. As her mind raced, looking for a winning strategy, considering every opportunity and possibility she could exploit thrice, but even the best of her plans were extremely risky at best. She was then suddenly taken out of her concentration by Arlen who said, "...Mynedrie! Come, I know where we can hide." The two then sprinted through the back alley to find themselves at the back of a bar. Arlen took a little twig that lay behind a trashcan and used it to poke into an empty space between two stones just behind the rain pipe.

He moved around the space before dragging it back out together with a small door key. He placed it into the door and opened it. He then told her, "Inside, hurry!"

Once inside Arlen closed the door again with his key before stepping away from the door. They then both stood staring at it while the sound of soldiers rushing to kill or capture them were stomping around, echoing through the door. Even with the wall between them and the corrupted militia's orchestra of footsteps sounded loud and almost crisp through the wall. The fear of their enemies storming through the door was torturous to Mynedrie's and Arlen's nerves, with Mynedrie standing ready to shoot the first man that came through that door. Lucky enough however they heard the steps becoming more and more vague, until they no longer could be heard by either of them. Making them breathe a huge sigh of relief.

"That was close," Arlen said.

"It sure was," Mynedrie answered. "How did you know that key was there?" she continued.

"This bar belongs to my aunt. I used to hang around here as a kid and the family would celebrate almost all holidays here," Arlen explained.

Mynedrie walked to one of the windows and slightly pulled away the curtain that hung in front of it. Outside she saw only an empty, ravaged, stone street, filled with stuff as people started to flee for their lives and partly broken down market stalls that sold caught fish and sauces to go along with them, behind dark, grey, large, bricked buildings with windows broken and doors forced open in an attempt to find more steel and other metals to loot.

She pulled back from the window and on top of the stairs that was standing in the left corner of the bar stood a woman big in size and shape just fitting through the hatch just behind her. She threw a knife towards Mynedrie, who quickly dodged the incoming blade. Mynedrie took cover behind a table she had knocked on its side. She was about to counterattack until Arlen started to stand in between the two.

"Woah! Woah! Woah! Both of you please calm down before someone gets hurt!" he said worried.

The older overweight woman dropped her knives and said, "Arly!?"

To which Myna confusingly said, "Arly?"

The woman walked down the staircase to hug Arlen firmly

while still clenching several knives in her hands. The woman was shorter than Arlen, two whole heads shorter than him in fact, "Ow, how are you doing?" she asked while hugging him before looking him straight in his eyes and asked, "What are you doing? Have you not seen what those militiamen are doing outside?"

"It is a bit hard to explain, I'm afraid."

She then moved her head and looked at Mynedrie with suspicion and asked, "And who might she be?!"

"I am Mynedrie Kianistra, a mageguard who is here to stop the necromancer Drumir," Mynedrie said with a harsh tone.

"Your friend here is a bit full of herself, is she not?" the woman said.

"Full of myself! You threw a knife at me!" Myna said.

"That was a warning knife!" she said, before once more turning back to Arlen and continued, "I swear girls these days, back in my day when someone threw a knife at you it was seen as a compliment."

Arlen moved away from his aunt and stood once more in between the two ladies, "Myna, let me introduce to you my aunt, Anna. She is the owner of this... fine establishment. Aunt Anna, this is Mynedrie, she pretty much already introduced herself to you," he explained.

"What are you guys doing here?" Aunt Anna asked.

"I saw a group of corrupted militiamen pillaging the harbor and wanted to stop them," Mynedrie said.

"Seemed to be going great," Aunt Anna reacted.

"Why did you not leave like the rest of the people?" Arlen asked, trying to change the subject.

"Arly, dear, I have been running this bar for over thirty-five years and I am not about to let some 'corrupted militiamen' take it away from me without a fight!"

"Not like we can just get out of here anyway with the militia blocking the road," Mynedrie said.

"Yes, they have blocked the road, but not the river. If we can get to the river and get a boat, we might be able to escape to the fortress island of the yet uncorrupted militia," Arlen explained.

"How are you so certain that they are not corrupted as well?" Mynedrie asked.

"I checked it while you were still asleep after I found you with my smaller telescope. The men that were on patrol at the time seemed normal to me," Arlen explained.

"Well now all we need is a boat and a paddle," Mynedrie said.

"Well that is easy, there is a small sand beach right next to the docks. It always has at least some boats laying on it," Aunt Anna said.

"Then we just have to carefully sneak around to get one," Arlen said.

"I think it might be better if you stay here with your aunt," Mynedrie said.

"What, why?" Arlen asked.

"Well unlike me you don't have any training in stealth, or

any form of combat. So I think it might be better if you remain here with your aunt," Mynedrie answered.

"Perhaps I'm not the best fighter, but I'm not some pushover either. Besides I know the infrastructure of this city better than anyone else and I know enough about strategy to understand the need of knowledge like that," Arlen argued back.

Just before Mynedrie could say something back, Aunt Anna intercepted her, "Don't even try, once he has started doing something there is little you can do to stop him."

Mynedrie looked with wide eyes towards Arlen before saying, "Fine, just be careful out there."

Arlen and Mynedrie went through the back door to go outside. Before they went, however, Arlen turned to his aunt and asked, "Are you sure, you don't want to come with us?"

"Like I said before, my place is here. Do not worry Arly, I will be fine. I have traversed rougher storms than this," Aunt Anna answered.

They walked outside, carefully looking around to find out if there were no guards on lookout for them. They slowly made their way around the many small and thin roads of the harbor district.

They took cover behind a small wagon from where they could see the beach on which several boats lay on shore. Getting there however was going to be a struggle as there was around a hundred meter sprint between them and the boat. The road to which was filled with soldiers and most importantly Stria, who was giving out orders just in front of one of the main docks.

"We need a distraction," Mynedrie whispered.

Arlen looked around searching for something of use. He took out a knife that he kept inside of his pocket.

"What is that?" Mynedrie asked.

"Oh this, this is a retractable blade, safer to take around in your pocket and takes up less space as well," Arlen explained.

Arlen used the knife to loosen one of the stones that covered the road of the harbor's streets. With the stone now loose enough that he could take it out of the ground, he continued with the next phase of his plan which was to search for a target, that he found in the shape of a window, built inside of a house on the other side of the harbor around twenty-five meters away from where they were standing.

He got up from behind the cart and threw the rock, which flew through the air in an arc before hitting the glass and making it shatter into several sharp shards. He heard the shattering of the glass from behind the cart and he was not the only one, for Stria heard so as well. She and almost all her militiamen ran towards the building, where the glass broke, believing that they were there as well. As Stria and the guards stormed through the door however they would see no one inside. She thought that they were perhaps inside. However just before she gave the order to go and storm the floor above them, she noticed underneath the broken window something laying on the ground. It was the rock that

Arlen had thrown. She realised at that moment that she had been deceived.

She rushed back outside to catch them. It was too late however and she saw that Arlen and Mynedrie were already pushing a boat off the beach and rowing away. Further and further distancing themselves from the small beach and the harbor as a whole.

Chapter 9

They eventually came on the shore of a small island, on which the militia's base was built upon. Its walls were five meters high and without any wall towers, but there seemed to be at least one ballista per side of the wall. One was even aimed at their little boat.

When they came to shore they were welcomed by about ten men; some slowly approached with halberds, some with swords and shields, and some stayed in the back with crossbows ready to fire in case something were to happen.

Myna was readying her bow before Arlen stopped her and told her, "Perhaps a more diplomatic approach would be best before they stab us a million times." She complied, putting her bow away and stepping out of the boat, yet remaining tense and ready for any sudden moves.

The commanding officer, a broad and muscular man for his age, walked towards them. He stood about half a head taller than Arlen, who was himself slightly taller than average and a full head taller than Mynedrie who, by elven standards, was relatively small. His face was partly covered by a greying beard which was well kept, as was expected of someone of his rank, and almost covered his mouth until he spoke, "Arlen, I had not expected you here, and you brought someone. Who is she?!"

Mynedrie felt insulted by his blatant disregard for her. Before she could speak out about it, however, Arlen stepped in, knowing Commander Krutz well enough to not let him get aggravated in any way, "This is Mynedrie Kianistra of the elven realms. She is here for an urgent hunt against the powerful necromancer Drumir."

When Drumir's name left Arlen's lips, it seemed to go through the very souls of the soldiers present, sending chills through spines and murmuring among the soldiers. Except for Commander Krutz who unwaveringly asked the next question, "Drumir. How did he get here?"

"We tried to transport him to a secure location, but he was

able to escape during the journey, nearly sinking the ship in the process," Mynedrie answered, wanting to be the one to speak to him this time.

"So his escape. His capture is your responsibility," Commander Krutz said to Mynedrie in a stoic yet condescending tone.

"His escape? No. His capture, yes," she said, determined yet controlled.

"Then why are you here?!" Commander Krutz asked barely hiding his contempt for her as he spoke, squinting his eyes and moving slightly forward.

"We heard commotion coming from the harbor and went to investigate. There we found a group of, I guess it were around twenty-five men who were pillaging the harbor, but it was not gold or silver they were after; it was metals like steel, iron, copper and bronze. When the fighting started, there were suddenly several crossbowmen and even some archers who started shooting us from the rooftops. Luckily we were able to hide by my aunt's bar before we took our chances and sprinted towards one of the boats," Arlen explained, trying to hush it between the two of them before they would find themselves again being bombarded with arrows.

"I know all of that!" the Commander remarked.

"You do?" Arlen asked confused.

"Yes, you created those monoculars for us, remember? We saw the whole thing happen. What I want to know is why she was talking to that grey-skinned lady!"

"She is my niece," Mynedrie admitted.

"Your cousin. How convenient," the Commander said, as if all his suspicions were true.

Before he could say anything else, however, Arlen interrupted and said, "Commander Krutz, with all due respect, what you may be suspecting right now is wrong. In the short time that I got to know her she has shown to be of excellent character and an exceptional mageguard."

Commander Krutz looked at Arlen with a stern face that betrayed his frustration towards him. Seeing this Arlen continued on anyway, “I mean consider this: she slew that berserking beast that could have slain everyone on that side of the city walls had she not intervened. I mean what reason would she have to do that unless she had genuine intentions to help us? She even fought her niece in the harbor.”

“It could have been a trick, to get us to let our guard down,” Commander Krutz argued back.

“Then let her prove herself, send her on a mission to show her loyalty,” Arlen argued.

Commander Krutz looked silently at both of them before saying, “Come with me, now!”

They did as they were told and moved right behind Commander Krutz, being further surrounded by several of his militiamen.

They made their way through the militia’s gate, which were two dark wooden doors reinforced with dark, coarse metal.

Unlike the city's walls and the castle walls, its gate did not have a gatehouse and was only surrounded by bricklaying that was from the ground up made with the bricks being placed horizontal, but as they followed the shape of the gate they would be placed in a more and more vertical position.

Behind the wall was, unlike the castle whose ground was filled with tiles and street bricks and the occasional plant, the inner plaza of the militia base was a grey-brown soil with an occasional patch of grass. The plaza was roughly divided into several practice grounds, where soldiers were practicing their disciplines, a smithy, where a smith was busy repairing some broken weapons and a carpenter shop, where a couple of carpenters were just like the smith busy with not just repairing weapons but also the shafts of spears and halberds and making shields which had a layer of cloth on which was painted hastily the coat of arms of the Drune royal family, a black anchor behind a dark yellow background.

At the training grounds meanwhile were divided into three

parts: the first was an archery range where soldiers were shooting with recurved bows that were from tip to tip slightly longer than their arms and crossbows that were more compact in size and ammunition, yet from the same distance seemed to pack almost as great a punch. Next to that was a big circle that was separated from the rest of the base's courtyard with a simple wooden fence, in which two militiamen were sparring. Both being armed with sword and shield, slashing, clashing and stabbing with them like there was no tomorrow all while an instructor yelled and explained what had to be done. Lastly was the part of the training grounds that was reserved for spear and halberd training, practicing individually how to stab with them against dummies made of straw while wearing a foreign design military helmet and an angry face having been drawn on it with either coal or charcoal. They also practiced military drills as a unit, how to stand in formation and what moves to make in such a formation when given different orders.

The Commander took them into a building on the left side of the entrance, which was relatively small, only around five homes the size of Arlen's. Which was not nearly enough for a city militia that a city like Drune would need to both man the city walls and the castle walls and let's not forget the militia's fort itself.

The answer came when they moved through the reinforced doors of the building, walking inside they were met with a broad staircase, as broad as the room they were now standing in. As they moved through the room they noticed two doors on each side of the room. The left side from closest to furthest from the entrance were leading to the kitchen that mainly made the meals for soldiers and more refined palatable courses for officers and a storage space, so militiamen could quickly get what they need during a siege, like weapons, bolts and arrows, rationed food and water and the like.

On the other side of the room where doors that now going

from furthest to closest from the entrance contained behind them a stairway to the roof, from where soldiers could shoot with bows, crossbows and even throw objects at besieging enemies who tried to enter the building should the first line of defense fail and lastly was the last door, behind which were just simply a couple of toilets. In case a soldier needed to go.

They walked down the staircase which went several meters down until they reached a floor again. In front of them stood a long broad hallway leading out of the room and containing on both walls in the beginning a wall that led to the barracks for the militiamen. Walking further into the hallway were again on both sides storage rooms: the one on the left for weapons, arrows and bolts, the one on the right for food and water. Walking even slightly further than that you would get the rooms for officers which were subdivided into even smaller rooms. Then at the end of the hallway you had the Command room, in the middle of which stood a big, round, oak-wooden table, on which lay a chart of the city and even the building plans of the most important buildings.

The Commander moved around the table until it stood between him and the recently arrived duo. He placed both his hands at the edge of the table before saying, "Leave us alone." His soldiers complied immediately and left the room in an orderly fashion. When the last militiaman left the room and closed the door, the Commander started speaking again, "Never! Speak against me when my men are near Arlen or I will make sure you pay for the consequences."

"I understand," Arlen said, with a stoic demeanor.

"Make sure that you do!" the Commander said, after which he moved back to a calmer and more neutral posture. He looked Mynedrie directly in the eyes and asked firm and direct, "How do we kill him?!"

"Not easily," she responded.

"What do you mean?" the Commander asked.

"Drumir has an amulet around his neck that will project a

forcefield against anything that moves too quickly towards him and will turn anything that is made of metal into rust, wood to rot and flesh and bone to wither away if you attempt to take it off of him. Believe me, people tried."

"And that is why you were transporting him to where?" the Commander asked.

"A remote island just away from the northern coast of the elven realm where he could be held, studied and perhaps even be destroyed by the mages there."

"And this niece of yours, how did she end up in all of this?"

"She worked as a ship maiden that happened to transport him at the same time. My guess would be some sort of manipulation, or otherwise deal she made with him."

"Do you know how she turned the way she did?"

"My guess is some sort of dark magic, necromancy most likely, but more detailed a explanation I do not have."

"And that beast you have slain, any idea where he got that from?" the Commander asked.

"He must have used some sort of necromancy as well," she answered.

"Then that explains away your assignment," the Commander exclaimed.

"Our assignment?" Mynedrie asked confused.

"Yes, ever since Drumir took over the castle and most of the city's militia we have been getting strange reports of 'huge' and 'massive rats' that have made the sewers their home, and I will send you alongside a group of men to check out if that is true and if possible capture one for study, which mister Arlen here would be more than qualified to undertake. Any questions?"

"Just one: how big of a group are we talking?"

"Five men total, all equipped with a shield and short sword," the Commander answered.

"Seems rather big a group. I mean, how big can the sewers be?" Mynedrie asked.

"You'll be surprised," the Commander answered.

Mynedrie and Arlen would make their way to the boat where the five men had already gathered for the trip. As the Commander had told, they were carrying swords shorter than they would normally be equipped with.

They all stepped into the boat and made their way towards a tunnel. Out of the tunnel a stream of water mixed with dirt streamed into the river. Lucky for the small team there was a walkway on each side of the tunnel, allowing them to walk into the tunnel without making their feet wet in the dirty, stinking water.

One for one they climbed up onto the platforms, Mynedrie, Arlen and two militiamen on the left and three others on the right; one of the militiamen on the right platform was Sergeant Otto, a fairly young man with locks of blond hair escaping from underneath his helmet and moving around his sky blue eyes.

When everyone had left the boats he gave the order to move forward and everyone complied. Mynedrie and Arlen walked behind the two militiamen and would almost immediately start a conversation, with Mynedrie saying in a hushed voice, "So why does the city have these huge tunnels running underneath it? It could never need tunnels this size to expel all the waste that those toilets and sinks get thrown through them?"

"True, but you are forgetting that a lot of them had to be built by people who are in need of space to work and do the occasional maintenance. Every brick and every centimeter of ground was laid and dug out of the ground. In a relatively short time I might add," Arlen responded.

"Really, how? You speak as if you were there."

"That is so because I was. I designed every part of this tunnel system and oversaw its construction and in a mere three years every house, even the humblest of them was connected to the sewer system."

"But how? How did you do it?"

"Well, I had some help from the prison. Every prisoner, no

matter their crime, was sent to work on one of several tunnels that were being built at the same time."

"Why would you use your prisoners for such a task? Were there no skilled craftsmen?"

"There were, but they were few in number after a massive plague hit the city. It is actually the reason the sewers were made in the first place. Before that, people would throw their... stuff right out of the window."

"I had heard about something like that as a child. I was pleasantly surprised when that was not the case here."

They would eventually move left past a corner. There they would find something disturbing. Two oversized, black-grey hunched furs were standing over a carcass that had already been nearly stripped bare to the bone. Bones had been cracked to get to the bone marrow inside. What it may have been was unable to be told from where Mynedrie and Arlen were standing. What they were able to make out was the scratching and cutting of their teeth against the bones, the sound of which echoed through the brick-laid walls.

Just when Mynedrie passed the corner, the two small critters looked back with shining, red eyes that stared enviously towards them. The light of their torches having warned them of their presence.

The giant rats made a hissing sound towards them before trying to flee away towards a corner just a few steps away from them. Before both of them made it, however, Mynedrie was able to shoot her bow, whose arrow made of magical light seared through the tunnel, lighting it up as it went, hitting one of the mutated creatures right in the back.

The small group moved forward, while carefully listening to their surroundings. Two soldiers standing with their swords and shield ready for any sudden incursion as well as a soldier in the back standing with his sword and shield ready as well. The last two militiamen stood in between the two, with a torch in one hand and a sword in the other, with Mynedrie and Arlen standing behind them.

Standing at the dead, oversized rat Arlen would kneel to better study the creature. It was even more vulgar up close. Underneath its black-grey fur along the spine bumps started to protrude upwards from underneath the skin. Its teeth had become unbelievably long and sharp, with a dark yellowish hue over them. Even the paws had sharp claws on them that seemed they could tear through skin with ease. With its forward paws having more elongated digits as well which made Arlen feel an unpleasant sense of suffocation for some unexplainable reason.

"Eww, what is that?" one soldier said.

"A giant rat obviously?" another answered.

"I know what rats are! And this is no mere rat!" the first answered back.

"Enough!" the sergeant said, "We are here to exterminate, not to debate," he continued.

"Well, hopefully it aren't too many. I do not want to stay any longer in this stinking mess than I have to," the youngest looking private said.

Just as he finished saying that, a rumbling could not just be heard but felt shaking through the tunnels with the sound of squeaking and squirting becoming louder as well. Wondering what it was, the sergeant would look behind the corner and what he saw would make him yell, "BACK TO THE BOAT!" while pushing them to move. Arlen placed the dead vermin over his shoulder before running from who knows what.

Before they turned back around the corner they first turned, before finding the two rats. Mynedrie would turn and stand ready with her bow ready to shoot whatever came from behind the second corner. What she saw she was not prepared for however, as a horde of overgrown rats, that function more as a wave came pouring into the tunnels. She shot at it with just one light-arrow before turning and running like the rest of the group that had already reached the boat and was ready to depart from the sewer entrance.

“Wait, where is Myna?!” Arlen asked with worry to the rest of the group that was already trying to move the boat away.

“Probably mouse meat by now!” the sergeant said back.

Arlen stood up to attempt to get a better look inside of the tunnel, hoping to get a glimpse of Mynedrie running towards him, “Wait, look there she is!” he said while pointing at her.

Mynedrie was running as fast as she could. Faster than a human could on average but even then the horde seemed to only get closer and closer behind her. As soon as she stepped outside she jumped towards the boat, crashing against Arlen and his new dead pet.

With Mynedrie now reunited with the rest of the tiny crew, they immediately pushed the boat away from the tunnel entrance before the rats could pour over them.

Sitting in their boat only a few meters away however, they were all surprised that the rats showed extreme constraint and except for the forward few that were pushed outside by the hordes’ very momentum. Those rats that were pushed outside were nearly burning underneath the sun, their fur starting to smolder and their few patches of hairless skin starting to get burn marks, with smoke emitting from the rats like recently fired-up pieces of coal. Only the rats that had fallen into the water remained relatively unscathed.

That was not what was the weirdest about the whole scene however, as a hulking figure that barely fitted through the tunnel came forward and examined the tiny boat floating right in front of what it thought was his domain.

It slowly stepped further forward but still staying well behind the tunnel's entrance it grabbed onto the iron gate of the tunnel, its forearm covered by a dark-grey cloak which failed to hide the sheer size of its arm muscles. Its biceps alone were bigger than an adult man's head. Before it would try to close the gate it stood still. Before it spoke with a deep yet screeching voice, "You will not come back to our home... 'cause soon, your homes will be ours!"

It then slowly closed the gate shut before walking back into the tunnels with its horde staying around him at all times. The only rats that remained were the ones that were still outside after he closed the gate, their big bodies being unable to get through the rails of the gate with their big bodies, having the choice of staying in the sun or getting in the water. To prevent themselves from burning.

"Hey, hey! Wait a moment! We have to row back!" Arlen said to the rest.

"What, why?" one of the soldiers asked, perplexed.

"'Cause I need one of those rats alive for testing. Several if possible."

"There is no way we are going to be able to do that," another soldier almost yelled at him.

"Do you want to get rid of them or not?" Arlen answered back.

Begrudgingly they rowed back towards the small group of rats trying to squirm their way back into the sewer tunnels before getting back into the water to prevent getting turned into a pile of ash and bone.

They grabbed two of the closest of the rats that had a relatively low amount of burns, having mostly laid in the water of the river, sticking close to the sewers' entrance as closer you got to the middle of the river the harder its flow got and started pulling you with it, just as it had done with several rats already.

Arlen turned his expensive looking coat into an improvised bag with which he was able to fish out one of the vast amount of rats that were swimming in the river's waters, fighting against the stream while also occasionally submerging themselves to wet their head to prevent burning.

The rat he captured started to violently move, squirm, scratch and generally do everything to escape. To prevent it from breaking loose, Arlen instructed one of the men to sit and hold it in place, which proved a challenge for the soldier who was surprised at the rat's strength which seemed comparable to that of a cat. With their new catch secured the rest started to row as fast as they could.

Chapter 10

Back at the militia headquarters, Myna was walking to the room where Arlen was busy examining the creature's reaction to different substances. She was carrying with her two plates of food with some bread and a stew. While she was walking down the hallway she noticed that the sun was almost dark.

She walked into the room and saw that Arlen was mumbling to himself while the enraged animal was trying to scratch from inside his cage that was placed on a corner table on the other side of the room. Arlen, however, was too much sunken in thought to even notice the screeching vermin.

“Am I interrupting?” Myna asked, keeping a plate with a bowl of steaming stew and a piece of bread in front of his face.

“Not at all,” he said, smiling back at her.

Mynedrie took a seat next to him and asked, “And, any progress?”

"Not really, except if you count the fact that he doesn't like cages," he answered while throwing a large blanket over the cage.

"Anything I can help with?" Mynedrie asked.

"Well unless you know how to make an extra venomous rat poison we'll still be stuck here for a while."

"Actually... I think I could."

"Wait, really?"

"Yeah, my mother was an alchemist so I learned a couple of tricks and my father showed me which plants are poisonous and how to make poisons and potions with them," Mynedrie said, "but I doubt you can get them fooled into eating it," she continued.

"Well they don't need to eat it necessarily, as I am trying to make a liquid with fumes so strong that it forces the rats out of the sewers."

"So you basically want to make the sewers totally unbearable, even for them."

"Not an easy task I know."

"But not impossible," Mynedrie added, "I know an old recipe for a liquid we use in pretty simple stink bombs that can make you nearly cough your whole life out just by the fumes. I can probably find all necessary ingredients inside the city," she added.

As the sun set and the night had come they could hear loud screeches coming from outside. Myna and Arlen looked outside and saw many of the sewer creatures go outside of the sewers and into the city. Specifically east and west of the castle of the city where the rich merchants, landowners and government officials lived, with Arlen's house in particular sitting just east of the city

And the militia commander's family, who lived before fleeing to the eastern half. There were also many in the long street of houses where mainly the salesmen, militia officers and craftsmen lived. Mainly the squad leaders and their majors lived there with their families. Most of whom had also fled into the militia's headquarters.

The one with the least number of rats was the city district on the other side of the river which was also the poorest. It was being attacked far less than the other parts of the city for its sewers were far smaller than that of the rest of the city, with many of the militiamen living or having family living there.

There was a heavy discussion going on among the militia's leadership on which part of the city to go to and aid. After some insight from Arlen and Mynedrie, they decided to go to the northern part of the district to get rid of the creatures there and perhaps even control over the region, allowing Mynedrie to go with a few squads to go and fight off these sewer creatures, while Arlen stayed at the island base to continue the "fluid" creation process.

On the boat that was bringing her to the northern part of the city she asked herself against how many creatures she may or may not face. According to Arlen, it could be between 250 and 500 of these creatures in the northern part of the city alone.

When they got on land, however, she pushed the thought away from her mind. As they landed on the shore from there she could hear the screeching of the mutated animals and the screaming of people and livestock.

Only a fraction of the city's militia was still under the commander's control which was about 500 troops and much of those soldiers had to stay in the headquarters to defend it. Therefore there would only be 50 militia members going out to fight against these creatures.

The plan was to divide into five groups of ten, with Mynedrie being one extra, making her group consist of eleven people. The first three would go and hunt the creatures in the streets of the district, while the fourth group would patrol the coast for any mutated rats walking around or trying to swim across. The fifth group, with Mynedrie in it, would go to the only bridge that connected the district with the rest of the city. There, they would place some explosives that Arlen was able to make with some stuff he could scrape together. They would then also destroy the sewer line that went right underneath the bridge.

They placed the bombs right under the bridge. While they were doing that, they were quickly attacked by a group of mutated rats, which were quickly defeated by the militiamen, with a few of the creatures running back to the market.

Mynedrie realized that they would probably get most of the other rats their way. They lit the fuse and got off the bridge and took cover. As they waited, they could hear those rats coming to them through not only the streets but also the sewers as well.

They waited and waited for the bomb to go off while those abominations got closer and closer. For a moment, some of the militiamen started to question if the bomb would actually work, just as they were done talking to each other, the bombs exploded, right in front of the rats.

It made a whole piece of the bridge fall into the water onto the sewer pipeline, which, under the weight of the collapsed pieces of bridge, crumbled. However, the delay there was between the destruction of the bridge and the sewer pipeline allowed some sewer creatures to make it to the militia's side. This allowed them to surprise attack the soldiers who were cleansing the city of those abominations. They tried to drag some of the soldiers back into the sewers or just downright kill them until some brave civilians armed with anything that they had lying around in their home saved the militiamen before any could be injured.

The civilians who came to their aid then went with them to help against the rats who tried to cross the river by swimming across it. The current, however, made it quite hard for them, allowing only a few to get across. After attempting to get across the water for over an hour, the big one with a muscled body, thick grey fur, and big razor-sharp teeth ordered them to retreat.

All the rats went back to the other side of the city, which

allowed the militia's forces to reorganize their defenses. First, they barricaded the homes along the river into a makeshift wall with the doors and windows closed and reinforced in a varying number of ways. In the western part of their part of the city was a road and an open space. To defend that they used wagons, tables, and other stuff made of wood to make a strong and sturdy barricade.

During the creatures' first assault, they were able to wound several militiamen and a civilian who fought alongside them. Mynedrie accompanied them to the militia's base so they could get medical treatment. She went also to the base to tell the commander about the situation that they were dealing with.

She started by telling him about the huge rat that seemed to be the leader and that he could command them far more precisely than they thought at first, and come up with at least simple military strategies.

The commander then asked her if she saw any weaknesses

that they could exploit. Mynedrie explained that he seemed to be a coward for always staying in the back of the fight instead of the front, even if it could have helped him win the battle they had just fought.

He then asked about the civilian who was brought in. Mynedrie started to explain that during the battle, some civilians came to help them against the rats, saving some of the soldiers' lives. They continued to help throughout the rest of the battle and also assisted in making barricades. Some were former militia members, while others were concerned citizens, most of whom wanted to volunteer to provide further aid in freeing their city.

While the commander showed hesitancy, he did allow them to help defend that part of the city. He also asked her to let all veterans among the volunteers come to the militia's base to receive equipment.

Chapter 11

The militia stayed prepared all night for a counterattack from the sewer creatures yet none came, which meant that it was their turn to strike. The militia was busy preparing to attack the sewer station.

In the meantime, the militia also started to train volunteer civilians into the fight, and the veterans had been put into their own unit. While they were older than most of the other militiamen, they were all experienced and had been part of other military missions, with more than enough scars to prove it.

Mynedrie gave the young recruits advice on improving their aim. Arlen on the other hand was preparing the poisonous liquid to be transported onto a small boat. It was divided into three different containers; if one container happened to be lost through one way or another, there would probably still be enough to spread enough of its toxic fumes through the sewer system to rid it of the sewer creatures. Still, this whole mission was based on a few lab experiments. There was too little for them not to attempt it, however, for their current state did not grant them the luxury of losing the initiative, as the commander of the militia told him.

The militia put together a group of 20 militiamen who would help Mynedrie and Arlen in getting the liquid in the sewer station and even getting them in there, which would not be too hard as the station was only lightly defended for Drumir had no clue how important the place really was.

They quickly made their way into the sewer station through the front gate and into the front door directly towards the main pump. It was easy getting in with there being only five corrupted guardsmen and a few sewer creatures defending it. They connected the vat full of the liquid with the main pump. Mynedrie and Arlen walked outside of the building, looking at the sewer entrance in which they had entered the sewers for the first time. For a moment they both stared at the entrance in anticipation.

Mynedrie, being impatient, looked at Arlen and asked, "I don't get it, why isn't anything happening?"

Arlen stopped her talking by putting his finger on her lips without even looking at her. He then asked her with his attention still on the sewer entrance, "Do you hear that?"

Now that he mentioned it she heard indeed something, a rumbling coming from the sewers. As she looked at the entrance she looked with shock at the number of ratlike creatures coming out of it trampling over each other to get out, away from the toxic liquid, only to be greeted by the painful raise of the sun. They all immediately got in the river's water, which stopped the burning for a while but as soon as they dried up the sun's light would start to burn them again, which meant that the creatures were forced to stay in the water constantly swimming against the river's flow to not be dragged toward the open sea. Which meant that the creatures had the choice of either burning or drowning.

Then came the big rat running out of the sewer with a dirty

cloak around him protecting him from the sun, swimming directly towards the river bank and going directly towards the castle, where he thought he would be safe.

Mynedrie started to go after him shooting with her bow to take him down. One such arrow hit his foot which made him choose to take shelter in the market where his wound could heal for there it was outside of the sun.

Myna went after him into the market. There it was now really quiet filled with abandoned shops ransacked by the rats last night. There was very little of worth left there. With her knives ready she looked around, unsure where he would be coming from.

She then made a quick dash backwards dodging his surprise attack from above.

In a gritty tone the giant rat creature started talking to her, "You kill my ratlings! You die, now!"

Mynedrie stood defensive ready to fight the giant rat brute with everything she got.

She dodged the first slashes from his knife-sized claws. He was big but slow, not as slow as she thought he would have been with his size but still slow. That was until she saw a sharp piece of metal the size of a sword speeding towards her. She used both her knives to dodge the blow, only for the rat creature to use the momentum to push her against a wall. It then pushed away the sword-like object and tried to stab her again, but Mynedrie dodged it just in time, with it piercing through a wall instead of her.

She had made sure she had some distance between her and the creature, so she would not get surprised again.

She could see that this blade he was using was attached to his gargantuan tail. It was now stuck deep in the wall costing the creature much effort, so much in fact that he needed to use his hands to get it out.

She had an idea but it would probably be hard to pull off. As the giant rat came closer to her she kept going backwards, and the rat kept following her. It started to speed up and so did Mynedrie who seemed to keep an almost precise distance from him. Sick of this game the rat started to use his tail sword which had a much greater reach than his claws. She kept dodging the blows time and again until she got placed between a wooden pillar and her enemy. The rat thrust his blade at her with full force; she however was able to once more dodge it, this time by jumping upwards and holding herself to the railing above. The rat tried to grab her with his claw, Mynedrie however was able to get out of reach by 'throwing' herself further away and grabbing the railway again. She then dropped herself back down. She took her bow and shot with it several times; the giant rat hid his face with one of his claws.

After all the shots were fired he revealed his face and let out a snarky laughter. It then tried to take its tail out of the pillar. Its first attempts failed so it wanted to once again try to get it out with the help of its claws. But then it saw what was going on; it saw that she did hit her target, the pillar. It then looked upwards and saw that cracks had formed around the pillar on the ceiling above.

He looked at Mynedrie who was aiming her bow, ready to shoot. Only she wasn't aiming at the giant rat but at the curtains above that covered the windows that let in sunlight during the day. She specifically aimed at the hinges that kept the curtains hanging on the ceiling.

With one well placed shot the curtains fell down to the ground, letting in the sun's light which started to shine upon the rat who in turn started to let out screeches of discomfort. Desperate it tried to block the sunlight with its cloak.

It started to fold up into one ball of fur and muscle next to the pillar in an attempt to fit entirely behind his mantle. Just as it thought it was safe and secure from the sunlight, it started to get attacked by Mynedrie, who used her bow to shoot holes into the cloak.

Sunlight started shining through the ever increasing number of holes, and that he was getting hit by her magic arrows did not help either.

Desperate to get out of the situation he grabbed a hold of his tail just underneath where the sword was placed; he first tried to take it off his tail by loosening the hinges that kept it together. When that did not work, however, he decided on a more radical solution which consisted of him biting off a part of his own tail.

Mynedrie was shocked by this sudden act of self-cannibalism. So shocked in fact that it took her a moment to realise that her plan had been compromised. Quickly she tried to continue her efforts by repeatedly shooting magic arrows at him.

Just as the rat was about to take its last bite from its own tail Mynedrie realised that the ceiling above the creature was already about to break apart so she put together all her strength in one, big finale overcharged light arrow to take the ceiling down as the giant rat took its final bite from itself and Mynedrie had shot her shot. As the monster was turning around and about to storm towards Mynedrie's position, he was engulfed in rubble.

However, the giant rat was not entirely buried underneath the rubble; its head and its giant arms were free from the ceiling's weight. This however also meant that they were exposed to the sun; with light emanating on its skin, the beast started to screech in pain as its skin started to show burn marks.

Desperate to get away the rat started to use its upper body strength to push itself and the rubble above it upwards. The rat was succeeding in pushing the rubble away with its strength, until Myna shot several volleys at the left arm's elbow of the creature which caused it to collapse on the weight it was trying to lift.

Back under the rubble the creature started to screech again. Mynedrie started to get sick of the screeching, she took her bow and shot a well placed shot in its skull. As she shot the creature, however, several of Drumir's guards started to show up.

Mynedrie climbed up the railway above her before they could spot her. She then put on her hood and tried to get back to her allies that were still in the north of the city, but not before she took a look back and saw that two guards were checking the giant rat laying beneath the rubble.

Chapter 12

Gritz's mangled up body was brought forth to Drumir who was in the medical room of the castle. Drumir inspected Gritz's body and looked at it disappointed, stating, "You oversized rat, how many gifts did I not bestow onto you and you still failed me! No matter, I shall just have to find a new purpose for your corpse. Stria, you wouldn't happen to be interested in some rat features, are you?"

Stria looked at the beaten up corpse and responded, "I want to get stronger but I don't want to get stronger like that."

Drumir responded, "Oh well, your loss. Besides, I did not design your current improvement with this anyway, but it sure would have been... (laughs) interesting."

They were then interrupted by Gritz who caught up some blood which caught their attention. Gritz then let out a weak, sad moan.

Drumir walked to Gritz and placed his scrawny left hand on Gritz's muscular, abbed and bruised chest. He then exclaimed to the giant rat, "Well, well, well, it seems that Gritz is still useful beyond being a resource. In fact, I think we can spare some of our resources to HEHEEHEEEE upgrade him a little."

Drumir wrote a list and then gave it to Stria, telling her, "Here is a list of things I need and don't waste my time bringing it!" Stria accepted the task and went out for the stuff. As she closed the curtains she heard a severe scream coming from the tent that was clearly Gritz's voice.

After spending a while ordering the troops to bring her the stuff Drumir needed, a task that seemed way harder for them than it should have been on account of their suppressed intelligence, Stria looked at the pile of muscle and metal that they called Gritz, a parody of his former self. Large parts of his being replaced with metal, fused to his meat and bone by dark magic. Half of his snout was replaced by a metal mask with in the mask's eyehole laid a dark green crystal which functioned as both a control assurance for Drumir, as well as an eye for Gritz to look through and Drumir to spy through. His left claw had been replaced by a metallic one with a rolled-up chain within, which inside was attached to the hand, which he could detach from the rest of the arm and use it as a mace or a grappling hook.

Last was his tail, which was mostly still there save for the part that had been cut off. From there another metal object had been installed, or rather two, as the metallic part of the tail split into two separate tails altogether. On one was a small ballista; the other had a tang-like end with which it could grab objects and load the arrows that were stored on Gritz's back.

Drumir walked to Stria and asked, "So what do you think!?"

Stria responded, "I'm not sure." On one hand it was a harsh and offsetting sight seeing an already grotesque creature turned into a half-rat half-machine; on the other hand, however, she could not deny that Gritz had been turned into a powerful weapon, one that could come in very handy later on.

“This time, however, he shall not stalk the sewers, but the countryside instead. Killing those who dare come too close to the city and preventing the militia from getting any reinforcements whatsoever!”

“Wherever he shows up Mynedrie will be there as well. Let me go with him, we can get her and kill her finally!”

“No,” Drumir said.

“No, what do you mean no?!”

“I need you here to lead the troops and keep control of the population and he needs to make trouble for our foes in the countryside. Both of you need to get me as much time as possible for my project.”

Chapter 13

Mynedrie looked at her reflection in the small cup of soup that she had gotten for lunch. She saw how dark rings had completely taken over the underside of her eyes, which was no wonder considering that Gritz had kept her and the rest of her squad awake for over three nights by making his terrible, loud screeches or throwing the corpses of dead cows and sheep wherever they were camped. Two times even one of her own team that had been separated from the group was thrown their way, all covered in bite marks, as if he had fed himself with them. Doing all of this while staying out of the sight of the main party, but not Mynedrie's; to her, he would allow a glimpse to be shown to her and only her.

The people that lived in this area had long since fled for safety inside the city's walls. The parts controlled by the militia that is. They had described how it seemed to attack anyone and everything like a predator, but almost only at night; during the day it was hiding inside one of the houses or barns of one of its growing lists of victims.

For this reason the commander had put together a squad of four men with hunting experience and Mynedrie who had defeated Gritz already. But this time was different for Gritz was not just a giant rat anymore; he now had a huge metallic claw, which could detach itself and with its over twenty meter long chain was able to use his fist as a flail, all the while still being able to control the hand itself. When he has either struck, grabbed or just wants to strike again, he can retract his hand back by rolling up the chain back into his metal forearm.

They were sitting on the northern bank of the Drun River using it to clean and rest before continuing the hunt. She started to loosen her hair so she could wash it with the river's water. An end from her, there was one of the soldiers shaving himself, using the reflection on the water as a mirror. As he was done he looked at his reflection with a smile before that smile turned into shock. Before he had any time to react, a hand with razor sharp claws and fur almost all over except the palm of his hand grabbed him, the claws burying themselves inside his flesh. The soldier screamed in agony as his right shoulder and chest had been impaled by thc four claws. With one hard pull the arm tried to pull the soldier into the water only stopped by the quick shooting of Mynedrie, who shot him in the forearm making it let go of the young man, who then crawled away from the water, barely able to use his shoulder and chest muscles.

Mynedrie and the other soldiers went to the wounded individual, "This is bad! He needs to go to a doctor, or he is not going to make it!" one of the other soldiers remarked.

"He needs to be brought back to the city!" Mynedrie said.

"But how? That beast will easily pick us off if we have to carry him around!" another soldier argued.

Mynedrie thought about what she could possibly do about the current situation before saying, "You will carry him back to the city, meanwhile I will hold Gritz off for you. When you're back in Drun tell them to bring reinforcements!"

"Are you crazy? That thing will kill you!"

"Do not worry, I have beaten him before."

The soldier looked at his wounded comrade with worry before exclaiming, "Guess we don't have much choice."

Mynedrie helped him with bandaging the wounds as much as possible before they set off to return to the city, the healthy soldiers supporting the wounded soldier with walking, making slow progress together on their way back.

Mynedrie meanwhile walked the other direction as she had noticed that Gritz had swam back up the river by the trail of blood that had been created in the water. The trail was quickly fading away, but if she walked along the river's northern bed long enough then she could perhaps find a trail of footprints, or even blood that came from him.

It would not be long before she had found her trail, footprints and even blood covered the river bed in one straight line. She followed it through the countryside eventually leading to a small farmstead that had been ransacked by him before.

She slowly moved towards the homestead, ready to draw her bow if she saw any sign of Gritz. At the front door, which was wide open, she could see that it was dark inside and the table in the living room had been pushed aside.

She drew her bow, ready and quickly stormed inside, first checking the left side of the room before checking the right to ensure that Gritz was not in the building. She looked around. The living room was a complete mess, furniture, food and belongings were all scattered around like a force of nature had thrown it around. On both ends of the room were windows, both having wooden shutters to keep the warmth in at night yet both were open, letting in light and breeze into the room.

The wall underneath the left window was covered in several dried up streams that had created a pool of blood. On the windowsill were the traces of fingers creating tracks.

She looked outside of it looking at the grey, white clouds

hanging in the sky above, contrasting with the dark, plowed earth right next to the home. She saw how the wind was moving the leaves in the trees, it was one of those cold winds that announced the arrival of rain.

She stared towards the outside for a minute before she decided to continue on her hunt. She then without any warning heard a loud bang from up on the roof. She wanted to storm outside to find what it might be, only to already see that it was Gritz, who had used its big metal hand and claws as a flail, striking with it the roof of the building that got heavily damaged.

She could not go out of the door, as Gritz was standing just a few meters away from the entrance, ready to charge right towards her if she tried and staying inside was also not an option. So out of the window she jumped, she made just in time to get out of the building.

She then climbed up on the farmstead's hay roof on the top around the middle of the roof she would lay and shoot at

Gritz, who was surprised that she could hit him from even there, specifically hitting him several times in the right leg, injuring him severely.

While Gritz had become slower, he had also become more wrathful as well. With a loud screech he threw his flail claw again towards the roof, all while she kept shooting towards him and her arrows kept piercing his body left and right. All parts of it except his face which he had covered with his other arm to protect the crystal that was powering him up.

From above the flail-claw came hurrying towards her, the massive piece of metal hitting the spot she stood just mere seconds ago. She was aware that she would be unable to keep the high ground. She quickly had to come up with a plan. Gritz however seemed to have changed his plans as well, the sound of his giant feet, squishing in the black mud outside, coming ever closer towards the homestead.

With him slowly closing the distance, she finally came up with an idea: all she had to do was wait till he was close enough to the homestead until she could strike him from above and end this days long battle. Forever!

She waited and waited carefully using her hearing to figure out what he was doing exactly, hearing how the giant rat entered what was once the living room. Moving around debris, trying to find Mynedrie in between the rubble that lay all around the former living room. Clearly he had assumed that one swing had hit Mynedrie or at least taken her down with the ceiling as well.

Quietly she got up. Slowly moving towards the edge, peeking to look at Gritz, carefully moving to not be seen by him. She saw Gritz with his back towards her, looking through a pile of rubble, this is my chance! She thought.

She took one of her knives, jumped off the edge of the roof, ready to stab Gritz from behind with the knife that was clutched between her fists. She was close to an arm's length away from her prey, ready to strike him from the shadows. Then in the wink of an eye she was grabbed by Gritz's massive left claw. Grabbing her by her waist. The sudden change in speed and utter shock of being grabbed made her fill her head with images and ideas of what was coming next now she was in his clutches. She looked into Gritz's eyes, one was still his original eye, the other was replaced by a green, well cut crystal, which was powering the metal parts of his body. In those eyes she saw nothing but contempt, wrath, envy and utter hatred for her.

She felt the grip of his claw tighten around her waist making it feel as if her ribs were on the point of cracking. She raised the knife in her left hand upwards readying to stab his wrist to cut herself out of his grasp.

Before she could trust her blade, his hand moved. Mynedrie quickly gazed at Gritz's face seeing that his knife-sharp, spit covered teeth were exposed through a vile looking smile, ready to rip apart anything they bit into.

She quickly used her left arm to protect her head. The teeth bore themselves into her upper arm and shoulder's flesh, reaching all the way to the bone, “AACH!!!” she screamed in agony.

While the pain was unbearable she was able to keep her thoughts together, focusing to reach for her other knife with her right hand. Attempting to ignore the pain, at least enough to prevent herself from fainting away.

She felt how he increased the force on his bite, crushing bone and ripping through muscle before ripping the arm apart entirely from her, making her let out an even louder scream, “AAAACH!!!”

Just as Gritz started to triumphantly bite and swallow her arm, looking upwards as he did, she took this moment to cut through the wrist of the claw that was holding her captive, cutting his tendons as she did, making it turn numb. Gritz let out a terrible screech of pain, “EEECH!”

She quickly moved her blade upwards piercing through his neck, then making a cutting motion so deep it almost decapitated him. She quickly moved away from Gritz before he fell over, looking at her with burning hatred for her in its eyes for one last time, a massive bonfire that smoldered down as fast as his last breath left its body.

Mynedrie slowly went closer to the still laying carcass, poking it with her left foot. When it showed no reaction she moved to its head, where she proceeded to kick the green crystal. When that did not work she started stabbing it with all the force she could muster, making it shatter into multiple pieces.

She walked back into the house and took a small bottle with a transparent liquid inside and a roll of bandages. She opened the bottle and then started to pour its liquid all over the wound, making her try to hold back a scream. When the bottle was emptied she then took the bandage roll using it to cover the wound, binding the bandage together by using her right hand and jaw to knot together the bandage that was going around her upper torso and back, turning around both the right side of her neck and underneath the armpit of her right arm.

She then started to lay down on her right side on the dirty floor of the living room, as she did not want the already red turning bandages from becoming dirty. Despite the pain still calling for some of her focus she fell into a slumber, for the need of rest exceeded the pain's need for attention. Her dreams would not relieve her from any burden however as they would start haunting her mind as well, bringing her back to the day she first saw true horror that Drumir was truly capable of.

Chapter 14

It was little over two months since she had bested her cousin in their duel and around a month ago that she won her trusted bow, Heartnailer, in the grand provincial tournament. Yet both moments were remembered by her as if they had only happened yesterday.

As for today, however, she had no time to daydream as her father was taking her into the swamps that her village lay beside. Most would see it as no more than a calm, wet area filled with the sound of occasional bird chirping, with a strong, unpleasant smell hanging in the air.

Yet in reality it was filled with huge beasts, potent plants and provided isolation from the rest of the elven realm, both its troubles and its aid. Making it a destination beloved by practitioners of the darker, magical disciplines, especially those who are particularly foul.

It was why unlike many other Mageguards who were mostly sent on missions all around the realm and even on occasion to foreign lands, her father was mostly just busy in the swamps. While he did not outright say it, she could see in her father's eyes that he missed traveling as often as he did. Still he had plenty of adventures to tell from this swamp alone.

They were making their way towards the horse stable next to the swamp gate of the eastern part of town. Looking onto the building there was a small and calm river which divided the stalls from the so-called heat district, which is where all the jobs were being performed that required a substantial heat source, mainly smiths and charcoal burners.

Both areas were surrounded by the trees of the village, around which were the homes of the villagers built. With wooden structures connecting the more higher lying houses with each other. Most homes in the village were as it was three houses with only the homes of the smiths and the richer aristocrats being built on the soil of the earth. The surrounding area and building materials of the houses did differ greatly however; those of the aristocrats were built mainly of stone and more exotic woods while those of the rest of the village were made of wood of the surrounding forest.

As she was preparing her horse she saw an imperial bureaucrat speak to her father. She could not make out what exactly, but knew that something serious was happening by the fact that they were speaking in hushed voices to each other in a corner of the stables, right next to the entrance.

She walked towards them, with her fine steed right next to her. She saw her father turn his head towards her with an expression of concern he failed to hide. "Is there something wrong father?" she asked in the most formal [dominant elven language] she knew, hiding her accent as best she could, as was expected in the presence of an imperial official.

He turned fully towards her and said with a worry in his voice, "There is... I need you to go back home and stay there, at least till I'm back."

Mynedrie was shocked by this sudden turn of plans, "Why... if I may ask?" she asked while almost losing her composure. Before her father could give any answer he was interrupted by the official, "I'm afraid that is classified my dear girl," saying it with a tone of smug self-importance.

"Very well father," she said disappointed and went on to bring her horse back to its stable before she would go back home.

As she took her saddle off the horse she said, "Ah I'm sorry Chestnut, we will just have to go ride another time." The horse moved its head and made a neigh sound, as if it was protesting. "Oh don't exaggerate, we will just have to wait a little," she continued.

She moved back to the horse's head and petted it gently before saying, "Alright Chestnut, see you later."

As she walked through the town's upper, wooden, street level she rubbed her right hand's thumb over the string of her bow—a little habit which she lost after today.

She took the bow off her back and took a good look at it. She had it for only a month but she had already well developed her skill with it.

She had gained such mastery over the weapon in fact that she could almost be considered a full-fledged mageguard. After four years of hard training she would become what she had striven so hard for. She remembers how she fantasized how she would gain the blessing of a veteran mageguard, most preferably from her father. She then would go to the mageguards' headquarters which was in Alti'tu'ru, which translated to "ever-gardens rest," and is the main economic and cultural centre of the elven realm. The city is mainly made of wood, with any stone or metal work you find being used to properly regulate the flow of water that runs in the canals through which products and supplies were spread throughout.

That it was a morning like any other, her mother had woken up before them giving praises to the stars for bringing them light in the night. She worked as an alchemist in the alchemist gardens of her hometown.

Despite the regrettable setback, it seemed that the day was keeping its good weather, which was something you did not want when you were going to the swamp, as the heat and sunlight would let a flower start to blossom that stunk with such ferocity you almost choked on it.

Her father was not the only mageguard that patrolled the swamp; several others were there as well, mostly those who were just made into mageguards. For while the swamp was well guarded and full of dangers, it were exactly those same dangers that got people to try and get something interesting, usually the teeth or bones of a mighty beast, or a fascinating and outstanding plant.

However, it was Drumir himself that came to visit the swamp that night, looking for something more... mundane. Specifically a mushroom, Greyspots they were called. They are light grey of color with spots of darker grey spread throughout of the shroom's surface.

Its medicinal properties were little; it could be used to make a laxative. You would need quite a lot of these mushrooms to make enough laxative for one person—about 20 for a litre. Luckily Greyspots grow quickly and in the swamp they grow nearly all year round, except during the winter.

It was this ability to grow quickly that made them interesting to Drumir, as with some minor changes with advanced dark necromancy they could be turned into something far more destructive—something that Mynedrie would get to notice for herself soon enough.

She was about to turn back home when suddenly a massive amount of hisses and moaning could be heard from underneath them followed by screeching. She took her bow, ready to shoot whatever it came from; her father had done the same. They walked to the edge of the wooden platform they were walking on.

Beneath, she saw a massive crowd of grey, humanoid

looking creatures with dark grey spots spread throughout their bodies and also odd angular, rock-shaped growths that differed from person to person, if they could still be called that.

She saw that one of the creatures bit into the arm of a man, though afterwards it easily let go of him and the other creatures ignored the bitten man as well. This was not without reason, however, as the creatures had achieved their goal with him. Mynedrie did not know what it was yet but she would know soon enough.

Meanwhile she had a job to do, which was to safeguard the people of the town. She saw a woman and her baby child that were driven into a corner, the mother held her baby close to her and desperately tried to shield her child from the vile, encroaching, greyspotted menace.

Then seemingly out of nowhere the monster fell, only a singular, circular, smoldering burn at its left temple of its head giving a clue where her and her child's saviour struck from. Mynedrie shot several other of these creatures specifically to make a way to the stairs that led to the upper level.

Mynedrie went to the woman who was grateful for her aid. Mynedrie told her to stick with her to find a safe place for her and her child. The woman obliged without protest.

Just before she and the woman could run away, Duran followed by the imperial bureaucrat found them and rendezvoused with them.

Mynedrie, Duran, the official, and the mother with her child were making their way through the streets of the town, though at a pretty slow pace as these creatures were in their way and seemed to only get worse as they moved further into the town.

They would cross a bridge that had two support arches on top of it and that were made of wood. Trying to give themselves some breathing room, Duran took his magical heat axe and used it to cut apart the thick support arches with just a few good swings, making the bridge far less stable. Beneath them ran a thin, shallow, little river. Small enough that you could even walk through it. The mushroom creatures tried to cross over the bridge and even some over the small river through the mud, which slowed the already slow creatures almost to a standstill.

Despite the massive increase in weight of the horde and the damage that Duran had inflicted on it, the bridge structurally held itself in place.

Duran asked his daughter if she could shoot an explosive shot at the bridge to destroy it, slowing the horde down in the process.

Mynedrie looked at the little bridge with uncertainty, doubtful if she had the skill to do it. She had been able to do it before, but that was during training and she never had done it with so much pressure around her. Her doubt dissipated however once her father put his hand on her shoulder. Mynedrie looked up at him and she saw a warm, calming smile coming from his face and he said, “You got this.”

She walked to the edge, her heart filled with renewed confidence. She pointed her bow towards the bridge ready to shoot. She controlled her breath just as her father taught her, focusing the bow’s energy, then releasing the arrow towards the little bridge.

With an echoing sound the explosion tore the bridge apart, both small splinters and big pieces of broken planks and beams flew through the air. Wood was not the only thing that spread through the morning sky, however, with the official of their group even having to puke, for the infected that were on the bridge spread as well—chunks of them anyway, but no blood they noticed which surprised the whole group and worried them even more. As they were trying to figure out why there was no more blood in the bodies, Duran stated that he had noticed something odd as well, for he had seen that they were starting to cross the river now that the bridge was destroyed. This meant that these creatures had no real sense of self-preservation and instead were driven by a sense of purpose.

The contemplating had to wait however as they made their way to the great townhouse, which is the bureaucratic heart of the town. It is also the home of the town's mayor and liege, a library and archives and even the alchemic gardens where her mother worked.

It is also the strongest built building of the village, almost like a wooden version of the humans' castles. Such huge buildings made out of stone Mynedrie could not believe it was true even with most people telling her that they did.

They made it almost to the townhome and not a moment too soon as they saw from a road right of them that a huge swarm of infected had been trying to get on the town square but were slowed down by the city's guardsmen who kept their distance from the creatures and only used their bows to take them down.

They had slain many of this grey plague, the horde however

showed to be too numerous to defeat. The guards instead were trying to delay their advance so the civilians could retreat into the townhouse and barricade themselves in.

Mynedrie and her father fired their bows attempting to alleviate some of the burden the townguard was facing. Around twenty men and women of the townguard were armed with spears and were spread into a single file slowing down the horde as much as they could.

Despite their best efforts they were being pushed back by the grey creatures who seemed to have little in the way of self-preservation, pushing through the spear wall while being pierced by them. Archers tried to drain the strain from the spear wall by shooting as many of the greyspot-creatures as they could, but they seemed to make little change. Still the spearmen and women slowed their retreat as much as they could.

Slowly they moved closer to the double doors of the townhouse, holding them open until the last moment so everyone could find safety inside. With only a few more meters behind them the group of eighteen men and women dashed to the entrances before being slammed closed behind them.

After the townguard had made it inside it became increasingly apparent to Mynedrie that she could not find her mother. She immediately told her father, who immediately started to help find her and screaming her name, “Mayiaria, MAYIARIA! WHERE ARE YOU!”

“MOTHER! MOTHER, WHERE ARE YOU!” Mynedrie yelled alongside her father.

After having searched everywhere Duran came to the realization that she must still be in the town’s alchemy gardens, which were left of the townhouse. He quickly moved to the windows on the second floor of the building’s left side that gave an overview over the alchemy gardens.

And there she was pressed against a dronbontree on a little hill, with a small group of infected edging ever closer to her.

Her work dress ripped and sullied, while she pressed ever harder against the tree's brown-grey bark, fear nearly freezing her in place.

Desperate to save his wife, Duran broke the window, then he grabbed his grappling hook, swung it around a big tree branch, then swung towards another branch that was closer to the ground.

He then took out his heat axe, jumped down on top of the infected beneath him and continued to slam his axe into the head of the abhorrent creature. It then almost immediately burst into flames, almost burning Duran as well.

He immediately started sprinting towards his wife, who was only a few steps away from becoming a nice meal for the infected and join their ranks in the process, mere centimeters laying between her face and the fingers of the greyspot, and almost he could lay his teeth into her.

Mayiaria felt nothing however, only hearing a hard splashing noise, followed by the sizzling of water right afterwards. As she opened her eyes she saw her husband's magical heat axe stuck inside of the creature's skull.

The other infected changed their focus from Mayiaria to Duran who fought them off with mainly his bow, from such a short distance it was nearly impossible for him to miss, sometimes shooting them at nearly an arm's length away from him.

One for one he shot his enemies with his bow until none stood in the nearby vicinity anymore. He held out his hand to his distraught wife who took it legs shaking and a face filled with tears of fears that sprung from her leaf-green eyes and ran down her fair face.

Together they sprinted towards the wall of the townhouse. All the windows on the ground floor had been barricaded with whatever could be found inside. Mynedrie dropped the rope of her grappling hook so her mother could climb upwards.

"Impressive," someone suddenly exclaimed from behind

Duran's back. At first he thought that it was another civilian or maybe a member of the townguard, but something about that voice was just so... unnatural.

He looked behind himself and he saw a Greyspot creature unlike the others that surrounded it. It was almost two times the size of the others with bonelike, curved plates protruding from its skull, underneath that was a huge grin, which showed his long, sharp teeth.

Duran turned to the creature and asked with a serious tone, "Who?! Or rather what are you?!"

"My apologies, I seem to have forgotten my manners," the creature responded and immediately continued, "My name is Thumir, you may recognise the similarities with my creator's name, Drumir."

Mynedrie looked shocked at the conversation between Thumir and her father. She exclaimed under her breath, "Drumir! Not him! Please! Not him!"

Duran looked with contempt at Drumir's latest creation,

exclaiming to it, "Drumir is nothing more than a parasite in need of exterminating and you are just another one of the plagues he carries with him!"

"Intriguing speech! Tell me, did the inspiration come from your ignorance or your bravado?!" Thumir asked with a sense of mockery in his tone.

"I grow tired of your speeches, you monstrosity!" Duran responded before firing his 3 remaining arrows: two hit the Greyspot creatures in front of Thumir, while the third was meant for Thumir himself.

The arrow did not hit however, as it was blocked by the long and broad claws that protruded from its fingers, which it was able to move towards the arrow's trajectory with surprising speed from its long, slender and pale arms.

"Good attempt," Thumir exclaimed, then stepping over his fallen Greyspot pawns. Duran now saw how extremely slender Thumir's body was; his entire body was extreme—elongated legs, which seemed barely able to walk; arms even thinner than that, with the aforementioned claws at the end. His waist being about as broad as its two legs pressed tightly together, only his protruding, oval shaped ribcage made it slightly thicker at the upper part.

It deeply unsettled Duran who found it hard to look at. Thumir meanwhile proudly spread his arms out and said, "What is wrong mageguard? Can you not appreciate beauty?"

"Beauty is to the beholder and I behold none to you," Duran responded.

Thumir stretched his overly long left arm and used one of his claws to make a 'come closer' sign, while telling him, "Then perhaps you just need to take a closer look! Not that you won't seeing how you have just exhausted all your arrows!"

Duran did not want to admit it but he was right; he only had his heat axe left to fight Thumir, whose claws seemed dangerous enough from this distance. He quickly reminded himself that he did have more experience than this overgrown and arrogant fungus, so he should be able to outplay him in battle.

Duran took out his axe, he then told Thumir, "You should know that one of your pawns burned to a crisp when he got touched by this."

"Oh I know, I was there controlling it remember? That is how certain I am that you will not touch me with it. Well, except for my claws."

"You do love the sound of your own voice, do you not?!" Duran exclaimed.

Thumir responded back, saying, "Oh I certainly do. I also love it to use it as a distraction!"

"A distraction? For what?" Duran asked.

Thumir paused for a few seconds before responding, "...this."

Then suddenly a horde of Greyspots forced their entry into the townhouse attacking both townguard and civilians alike.

Duran screamed to Mynedrie that she had to keep the civilians and her mother safe from them.

She complied with her father's orders and started to shoot

the Greyspots from her elevated position and telling her mother to stick close to her.

From outside the screams of innocent men, women and children could be heard, together with the screams and clangs of battle from the townguard and the infected. Duran tried to ignore it and instead focus on the battle right in front of him.

Thumir spread out his right arm and started speaking, "Before we start this I must tell you, this. is. not! What you think it is. You are not a mighty beast-slaying hero and I am no mere beast... the truth is that I am a king and you a mere pawn."

Duran grew tired of Thumir's constant gloating, so without having much of a plan he decided to go for a direct attack. With heat-axe in hand he sprinted towards Thumir before making a jump so he could chop right into his face.

It did not work. Thumir slapped him with the back of his left claw against the tree his wife stood mere moments ago.

Thumir then raised his claw to stab Duran who evaded the attack, "He is tougher than he looks, but taking him down should not be too hard."

Then suddenly his left claw started to move towards Duran at incredible speed. He tried to evade this second attack as well but got hit by two of his claws in the right side of his stomach. Thumir then used his right arm to hold Duran, keeping him pressed against the tree's trunk while bringing him ever closer to his face, slowly opening its wide, grinning mouth, revealing just how many sharp teeth he truly had. Duran, completely stuck, knew he could do nothing but accept his fate with dignity.

Mynedrie looked with shock outside, seeing how Thumir was taking a bite out of her father's left side of his lower neck. For a moment she stood still unable to process what she just saw. She then immediately made her way to the window trying to go fight Thumir.

Thumir put a little distance between him and Duran. He

looked his injured opponent in his face and said, “Delicious!” He then used the claws on his right hand to block a shot coming from Mynedrie who angrily and tears rolling down her cheeks kept shooting at him. Thumir started to get it tough with the constant barrage of light arrows and so threw Duran away like he was nothing, making him collide against another tree.

Knowing that she would not get through Thumir’s defenses with just her bow she concentrated all her power in an explosive shot, which would be able to destroy Thumir. She would completely obliterate this crime against nature.

Thumir looked not impressed however. He seemed to stretch his arms out almost like he was mocking her, daring her to take the shot. Some of his pawns started to stand next to him making the same gesture as he was doing.

Totally enraged, she unleashed her now fully charged explosive arrow towards Thumir. However, Thumir had prepared himself, as soon as she had shot her arrow, he grabbed the closest pawn on his right side and threw it directly at the arrow's way, making them meet right in the middle between Mynedrie and Thumir.

Thumir then stated, "My turn!"

Mynedrie then noticed that all the Greyspot creatures in the Townhouse had lost their interest in anyone else and started to focus on getting to her. Mynedrie looked hopelessly around trying to come up with any sort of plan to get out of this situation.

"Face it young one, you lost! Soon the entirety of this village shall listen to me and my creator, soon the whole world shall be filled with his crea...tions..."

Suddenly Thumir could feel something stuck in his back, then a burning sensation that almost immediately became extremely hot. So much in fact that flames started to form on his back. The flames started to quickly spread across his pale body.

From up on the window Mynedrie saw that her father was in

fact still alive, though barely. With his last bit of strength he was able to throw his heat axe to Thumir, who had turned his back from him and towards Mynedrie.

Desperately Thumir tried to take it out from his back. His claws were able to take the axe out of his back, leaving a mark on the weapon's handle. It was too little too late however as the flames had already spread across his body.

With anguish he exclaimed, "No... no please not like this! Drumir please do not let me go down like this if only you allow me another chance... please?!"

The plea went unheard and Thumir burned to a crisp, with his passing his army started to turn on itself. A former somewhat cohesive horde reduced to nothing more than a bunch of brainless savages.

Mynedrie immediately went to her father, desperately trying to save his life. Her mother followed right behind her, burning her hands to the rope while using it to get down. Mynedrie used the same route her father had made. But she jumped to the rope he had thrown.

Her father was in bad shape however, if he did not die of shock or blood loss he would turn into one of those beasts if not after his passing, "I'm afraid it is too late for me Mynedrie."

"No... dad don't say that," Mynedrie responded.

"Duran! DURAN! No... by the stars no!" Myniaria wept.

"Ladies. Please... listen I do not have much time, if I do turn into one of those beasts I want you to put me out of my misery before I can hurt anyone innocent," Duran pleaded.

Both women had rivers of tears running over their face staying silent before Mynedrie broke that silence by saying, "Okay father, I will."

This was met by an expression of shock and disbelief, failing to express what was all going through her mind.

Chapter 15

Stria looked in disbelief at the little pool of water through which she could see from the perspective of the crystals of any of Drumir's lackeys that carried one. She gazed upon the scene unfolding before her, watching as Gritz bit off her niece's arm and how she retaliated with her knife. Drumir stood at her right, only saying, "All that work on him and still so little achieved, what a shame." He muttered disappointedly before turning away from the little pool, having seen her destroy Gritz's crystal, severing their connection.

Stria turned toward Drumir as he strode to exit the tent. "How... how could you let this happen?" she said, clenching her palms against the edges of the bowl of blurry water.

Drumir stopped, turned around, and asked, "What?"

"Mynedrie... you said I would be able to show that I am stronger than her, to regain my true status in this world, if she is unable to fight!"

"I have given you that chance; remember in the city mall where you let her escape from your grasp," Drumir responded.

"YOU KNOW FULL WELL THAT THAT WAS NOT THE DEAL! I WAS SUPPOSED TO FIGHT HER! I WAS SUPPOSED TO DEFEAT HER! I... I..." Stria found herself struggling to breathe. Her body, once strong, suddenly felt as though it carried a massive burden. Unable to keep standing, she fell to the ground. Her lungs struggling to inhale, her vision blurred, the sounds reaching her ears vague yet clear enough to make out what Drumir said.

"Yes, I did indeed tell you that I was going to make you regain your status, wealth, and dignity in this world. I did not say I would help you defeat your little niece in this rivalry you seem to believe in. In that, I gave you your chance, and you failed!" He kept watching her as she struggled for breath. All she could do was listen to him as he continued, "...as for the kingdom I told you you would receive, that you can still gain. But speak to me in such a manner however, and I will make your struggle with breathing last longer than a few seconds. Understood!?"

"Y-yes..." she struggled to say, barely able to speak.

After her clear response, Drumir released her from her discomfort and told her with a wild smirk beneath his crooked sharp pointed nose, "Good. Now go get some men and round up some townsfolk. Our opponents are growing stronger and more numerous and so must we."

Stria, still breathing heavily from gasping, said, "Yes... sir," before walking out of the tent.

When she had left, Drumir walked towards a sheet beneath which lay his latest creation, murmuring, "Shhhh, don't worry. As long as you do not fail me like that fool did just now, you will be alright."

Chapter 16

Mynedrie lay unconscious in a narrow hospital bed, pale and bruised, the steady rhythm of her breathing barely perceptible. Beside her, Arlen kept watch, his eyes heavy with worry and exhaustion. For the past three days, after the day's duties were done, he had spent every waking moment by her side, unable to offer more than his silent company.

Flashes of her battle against Gritz flickered behind her eyelids. Though Arlen could not see inside her troubled mind, the anguish manifested starkly on her face—contorted in discomfort as she muttered and grunted in her fitful sleep. Amid the cries and screams of pain, brief moments of silence punctuated the storm of agony.

Suddenly, she thrashed violently, claws clutching the bedsheets as if grasping for a lifeline. Arlen caught her by the shoulders, forcing her down before she toppled from the bed altogether, her cries piercing the quiet room, "Doktor!... Nurse!... Anyone help!"

Before aid could arrive, her eyes flew open, wild with panic, sweat pouring down her temple. Lost and disoriented, she gasped for breath.

Arlen's voice soothed the tempest. "Easy, Myna. Easy. You're safe. You're at the militia's headquarters."

Slowly, she calmed, and her voice mumbled, "Oh, Arlen, it's you. Sorry, I had a nightmare that I was back in my hometown and he..." Her gaze wandered to the bandaged stump where her left shoulder should be. Silent tears fell as the weight of her loss settled over her soul.

She spoke again, leaning back against the cold wall behind her bed. "It was no dream."

The militia commander entered then, pausing to regard her fragile state. Though hesitant, he knew it better to speak sooner rather than later.

"Miss Kianistra, on behalf of myself, the militia, and the civilians of Drune, I extend our sincerest gratitude for your continued aid in our struggle for the city. While you were not an official member of the militia, I hereby honor you with a metal plaque and a ceremonious helmet. I took the liberty of having the plaque affixed." He revealed the resplendent parade helmet carried by militia elites.

"Furthermore," he added, "you are now eligible for a standard militia member's pension and citizenship."

Mynedrie's eyes did not meet his. She murmured in a tone weighed with disbelief, "Thank you. I appreciate it."

The commander placed the helmet reverently on a nearby chair. "Well, I will leave you to it then." And with that, he departed.

Arlen's gaze lingered on the empty space where Mynedrie's shoulder once was. His eyes shifted to her face, mostly hidden beneath tangled brown strands, unkempt and loose. He reached out, placing his trembling hand over hers to steady her, but she snatched it away swiftly.

"Please, Mynedrie. Talk to me," he pleaded softly.

A crimson glare met him through the hair—tears trailing

down warm cheeks. “What is there to talk about? I’m an archer with one arm, a mageguard who can no longer safeguard others,” she whispered.

“I’m useless.”

“No, Myna, you are not. Look at what you’ve accomplished. Without you, there would be no army to oppose Drumir,” Arlen insisted.

“That was before I lost my arm,” she sighed, turning away. “Please, leave me to my thoughts,” she implored.

Arlen nodded, aching in his heart. “Very well. But know I am here and will help you.” He gave her a final look before leaving through the door.

He made his way to his workshop where Gritz’s mangled body lay for study. It had been brought back on the commander’s orders—for many reasons. First, to prepare for future battles against similar beasts; second, to bolster morale; and third, to prevent Drumir from wielding this creature again.

Before entering, Arlen knew he could help Mynedrie recover. He had already examined the limited materials Drumir had left.

He studied Gritz's metallic arm, particularly where flesh met cold steel. With a steady hand, he drew a scalpel and cut carefully around the adornments, intrigued by dark threads imbued into the nervous system. These 'dark nerves'—as Arlen named them—wove throughout the mechanisms, granting Gritz full control over the metal limb.

Taking scissors, he excised strands of the eerie material, noting its abundance. It resembled an infestation, yet it did not tangle in the moving parts.

Using a magnifying glass, he inspected further: dark, string-like necromancy with small outgrowths all over. He compared it to a similar sample extracted days before when Mynedrie and Gritz's body arrived, perfectly preserved in ice. Whether this lack of decay was miraculous or morbidly ironic was up for debate in Arlen's mind.

Mynedrie would never consent to using necromancy, but that was a hurdle he would face when the time came.

Moving to another table, he found a container chilled with ice beneath which lay Gritz's head, minus the crystal shards. Those fragments rested in a box hung on a wall—a next subject of study.

He examined the hollow spot where the crystal sat, noticing dark nerves clustered like glue holding it fast. Turning to the corpse, he pondered the intricate connection. The massive frame and metal augmentations demanded great energy, beyond what a mutated stomach could produce; so the crystal must serve as a potent power source.

Though sound in theory, proving it would be difficult, especially as the shards were powerless after shattering. There was a place, however—deep within the Drumir-controlled city—where magic crystals might be recharged. Dangerous, yes, but necessary if the create was to function.

He would have to wait until nightfall, lest the militia

commander forbid this risky venture. His duties kept him busy overseeing weapon production, giving him only a few hours to prepare the necessary blueprints.

Calculating the arm's weight roughly four kilograms, he selected materials accordingly.

Back at Gritz's corpse, he took steel fragments for crafting, weighing carefully against calibrated weights to ensure surplus.

Returning to his drawing board, Arlen labored over sketches depicting a prosthetic arm mirroring natural length and contour, blending seamlessly with armor-like enhancements. Where movement was crucial, supple leather concealed inner workings, including the palm.

Darkness fell swiftly and Arlen gathered his tools, the largest crystal shard, and samples of the 'dark nerves' into a bag. Grabbing a piece of fruit for sustenance, he settled in for a long, arduous night.

Chapter 17

Mynedrie lay motionless in a narrow hospital bed for several hours, neither sleeping nor turning, her gaze fixed blankly forward. Hunger, immense and insistent, crept upon her like a hungry shadow she'd tried to deny for days. It finally overpowered her will, stirring her from the bed after days without food.

She rose unsteadily, knife clenched tight in her right hand, and tried to open the door with her left—only to realize it did not respond. Looking down in disbelief, she saw no arm. Her mind snapped back to the missing limb on her left shoulder. Shock froze her for a moment, then she steadied herself and moved forward on her one good arm toward the kitchen.

The cold, dark stone halls of the militia headquarters enveloped her, sending shivers down her spine. She sought to warm herself as she traversed the dim corridors, the most meager of torchlight barely penetrating the shadows. In those shadows she thought she glimpsed cold, dark-red eyes staring venomously, lurking between flickering flames—a trick of the light or worse.

The kitchen was empty; night had cloaked the fort in silence. Only the pale moonlight betrayed the hour—nearly midnight—and all but the night-guards were asleep.

Mynedrie snatched an apple from the fruit basket, set it on the cutting board, and attempted to grasp it with her missing left hand. Frustration burned anew with the cruel absence of her limb.

Anger flared and she stabbed the apple with her knife, splitting it through the core before halting on the solid wood. Sighing softly, her blade not meant for such tasks, she proceeded to quarter the apple, arranging the pieces on a platter. A bite revealed the fruit to be bland and dry; not the luscious taste of home but the meager offerings of this northern village.

She pocketed the rest reluctantly and made her way back through the long, tunnel-like hallway. The torchlight barely dispelled the gloom. Mynedrie fancied she saw the sinister ratlike shadows skittering just beyond her sight, stalking and waiting.

Logic would deny their presence, but fear held her captive.

Slowly and cautiously, she hugged the wall—where torches hung—moving with deliberate stealth.

Along the walls, weapons hung in silent tribute: a sword, a shield bearing the coat of arms of Drune, an array of daggers—and a solitary bow.

The bow was simple, worn brown wood, its grain richly showing. Small and compact, it was a short-range tool meant for hunting, not war. Strange, she thought, to find such a shaft amid instruments of battle.

Lost in thought, the platter tilted—an apple slice slipped free and tumbled to the floor.

She reached again for it, inwardly grappling with the cruel truth of her lost limb. Repeatedly, she envisioned grasping the piece, only to find no response. More slices fell, mocking her helplessness.

Muscles aching from stooping, frustration spiraled into agony. She rose, eyes fixed on the scattered fruit—her failure made concrete and cruel.

With a scream of misery, she hurled the plate against the stone floor. It shattered into shards that scattered with deadly sharpness.

Exhausted, she sank to the cold wall, gazing at the ruin she'd made: apple slices mingled with shards of shattered pottery.

Alone, bereft, she reflected on how fate had brought her here, seated on the floor of the fortress, one arm lost to the belly of a great fallen beast.

Her mind spiraled endlessly through the torment—pain, loss, memories—that haunted her still.

The heavy footsteps approaching shattered the cycle.

Quickly, Mynedrie concealed herself behind the kitchen counter as Commander Krutz entered. He must have heard the plate's fall.

His steady gaze swept the dim kitchen, framed by moonlight and shadow. The fragments and fruit lay at his feet. Bowing, he inspected the scene but remained wary, scanning for threats.

Raising an apple slice, he regarded it quietly before stating, “I know it is you, Mynedrie.”

Emerging from her hiding, she asked softly, “How did you know?”

“My soldiers know better than to pilfer the kitchens, and no recruits could sneak past our night patrols. Only you, really, would venture out. And if you were a spy or assassin, you wouldn’t still be here,” Krutz replied.

“Thanks for the reminder,” she whispered.

“You fail to see what I mean. The men respect you for all you have done,” he said quietly.

Mynedrie seated herself beside the commander, voice low and bitter. “I won’t be much use to anyone like this,” she said, gesturing to where her arm once was.

The commander said nothing, simply joining her in silence. His gaze fixed forward, hers downward, despair a quiet intruder consuming them.

After a moment, Krutz began removing his armour's leg plate, rolling up his pant leg to reveal a prosthetic around his lower limb.

He stared blankly at it, his face tracing old wounds. "Lost it as a lieutenant, during the Trade War with Ozzburg. One of the last battles. Our forces were equal, save the trebuchet they set forth. A projectile crushed my leg."

Pride returned as he stood tall before her. "Yet even after that, I stand strong and proud."

"How?" Mynedrie whispered.

"I remembered I was needed. That I had value. That I had people to lead, and to teach."

Placing a firm hand on her remaining shoulder, his voice soft yet steel-edged, "And so do you."

This version enriches the somber mood and character depth

appropriate for a medieval fantasy healing narrative, presenting emotional detail within a gritty, loyal comradery setting.Mynedrie lay motionless in a narrow hospital bed for hours, neither sleeping nor turning, her gaze fixed blankly ahead. A growing hunger she had tried to deny finally forced her upright after days without food. Knife clenched in her right hand, she reached for the door with her left—only to realize that arm was gone. Shock washed over her, but she soon recalled why and steadied herself with her remaining limb, moving toward the kitchen.

The cold, dark stone halls of the militia headquarters sent shivers down her spine. Flickering torches barely pierced the shadows, where she fancied hateful red eyes glimpsed between the lights—though she knew better than to believe such fears.

The kitchen was empty at this late hour, the pale moon high in the night sky. She plucked a dull apple from the basket and placed it on a cutting board, again trying to grasp with her missing hand—frustration flaring anew. Stabbing the apple, she quartered the pieces and bit into one. The bland fruit was a far cry from those of her homeland.

As she retraced her way down the dim hallway, she pressed close to the wall where torches hung, moving with careful steps. Along the wall hung weapons—swords, a shield emblazoned with Drune's arms, daggers—and a simple hunting bow, its worn wood marking its humble purpose.

Lost in thought, she tipped the plate; apple slices fell one by one, mocking her helplessness. Frustration and grief boiled over; she hurled the plate to the floor, shards scattering like broken hopes. Slumping against the cold wall, she stared at the ruin she'd made—alone and broken, with one arm lost to a great fallen beast.

Her spiraling thoughts were broken by approaching footsteps. Mynedrie ducked behind the kitchen counter as Commander Krutz appeared. He must have heard the crash; inspecting the fragments with care, vigilant for danger, he said quietly, "I know it is you, Mynedrie."

She emerged cautiously. "How did you know?"

"My soldiers don't pilfer the kitchens, and no recruit could slip past the night patrols. Only you would be out. And if you were a spy or assassin, you would have fled or struck by now."

"Thank you for the reminder," she whispered.

"They respect you for all you've done," he said.

Together they sat in silence, her despair mirrored in his quiet gaze. After a time, the commander unstrapped his leg armour, revealing a stout prosthetic limb.

"Lost it as a lieutenant, during the Trade War against Ozzburg. A trebuchet's stone crushed this leg," he said, voice heavy with memory.

He stood tall, pride undimmed. "Yet even then, I stood strong and proud."

"How?" she asked faintly.

"I remembered I was needed. That I had value. That I had people to lead and to teach."

Placing a steady hand on her shoulder, he added softly, "And so do you."

Chapter 18

Arlen rowed quietly along the southern bank of the Druze River, shrouded in the inky cover of night. He knew well the risk—the threat of reprimand or even imprisonment for this trespass—but the weight of his mission settled heavy upon him. For healing Mynedrie, this secret venture was the only hope, and a task no man could undertake alone.

Tonight, he sought an old acquaintance versed in the mysteries of crystal magic—a figure who might cleanse the fragmented shards and purge the necromantic taint that clung to them like a curse. Yet it remained to be seen whether she would aid his desperate cause.

The boat slipped beneath the massive chains—vast iron links stretching from the mighty mall's edge to the western span of the militia's island stronghold. Each link dwarfed Arlen himself, its hollowed-out iron form an ingenious economy of precious ore. These chains, though simple, formed a formidable barrier against enemy ships and held fierce symbolic weight as the city's sentinel.

Yet to Arlen, these chains were a bitter emblem—significant yet powerless against Drumir's conquest, which had swallowed the city like a ravenous shadow. There was no time for brooding, for the mission pressed urgent.

Drumir's corrupted militia swarmed the western reaches of the southern city, lined along their river half, poised against any counterattack. Others garrisoned the castle seat, while scattered patrols guarded the labyrinthine streets.

Coming ashore north of the grand mall, Arlen faced its three-storied stone walls, windows and shutters ominously left open—an unusual sight, for shopkeepers once barred their wares securely at night.

No time lingered for reflection. Heavy bag of crystal shards, dark nerves, and mechanical arm slung over his shoulder, he slipped past the main thoroughfares, which pulsed with the watchful eyes of corrupted guards, into the twisting, shadowed marketplace where concealment was his ally.

At the market's edge, the temple of Drune rose—its pale stonework aglow beneath faint torchlight. Arlen timed his dash expertly, waiting for guards to turn and making for the open doors that surprisingly stood unlocked.

Inside, rows of benches held sleeping townsfolk, seeking sanctuary from Drumir's terror. A lone priestess presided over a deep stone bowl filled with liquid like the ocean's depths—a divine elixir, known for healing and comfort, gifted to the poor in times of need.

Though shrouded in darkness, the temple's marbled pillars, carved statues, and shimmering orb above reflected the elixir's pale blue light, casting a hushed, reverent glow.

Lost in thought, Arlen's reverie was broken by the priestess's voice. "Daydreaming again, Arlen?"

He smiled wryly. "Well, it is night, so technically no."

"That, it sure is," she said with a knowing smile.

Approaching, Arlen implored, "I truly need your help."

Her tone darkened. "And here I thought you came to visit, or at least join us for the night ceremony!"

Surprised, he asked, "You hold ceremony now, with all these people hiding from Drumir?"

"Do not speak that name in this sacred house," Sophia warned sharply, her voice echoing with sudden strength that stirred a nearby sleeper.

"I'm sorry, I did not mean disrespect," Arlen bowed.

She inhaled deeply and spoke calmly, "These are true believers, protected by holy wards. Abominations shall not enter while the ceremony lasts."

Curious, Arlen asked of the temple's defenses against the corruption.

"Remember Lord Matheus Aramea?" Sophia queried.

"Of course, but what has that to do with this?"

"We shall see," she replied enigmatically.

With new resolve, Arlen produced the dark nerves from his bag. Sophia gasped at their shriveled form when held to the elixir's light.

"What… are those? And why bring them here?"

"I call them dark nerves. Found them embedded in a metallic limb – one of Drumir's lackeys. They and other artifacts are why I came."

Sophia smiled, disbelief twinkling in her eyes. "…To purify them?"

"Something wrong?"

"Other than you asking me to do something borderline heretical? Not much."

"Sophia, you are the only one who holds such knowledge."

Her gaze swept the gathered faithful and rested on the awakened man. He grumbled humorously about unexplained screams.

"Very well," she said to Arlen. "We will need to move to the attic—there is more elixir stored for purity rites."

In that cramped loft, surrounded by faint candlelight and barrels of remedy, Sophia worked with focused grace. She combined tree sap and rare blue leaves, blending potions with the divine elixir until the blend shimmered a noble blue. Into this she dipped the corrupted shards, watching as the dark corruption bubbled and dissolved.

Clad in gloves, she prepared a thunderous elixir crackling with lightning. With careful hands she immersed the largest crystal shard. A few moments later it glowed with tranquil azure light.

"Take it," she urged, handing the radiant shard to Arlen.

Studying the purified crystal, Arlen dreamed of its potential—the lives his crafts might save.

Meanwhile, Sophia purified the dark nerves, which lightened from pitch black to pearly white.

Their work disturbed by footsteps—a hooded figure entered, cloaked and veiled.

A young priest, engrossed in ceremony, took no notice of this enigmatic visitor clad in hastily assembled dark robes.

The figure moved swiftly, casting a shadow over the bowl, extinguishing its pale light.

Alarmed, Arlen grabbed Sophia. “We must leave! Is there an escape?”

“There’s a hatch in the ceiling leading to the roof,” she replied, guiding him.

Outside the temple, corrupted guards rounded up terrified families in a circle of cold rain and steel.

Among them stood an elven woman of radiant steely skin, bobbed hair framing sharp features, her sabers marked with a tarnished coat of arms.

Arlen beckoned Sophia urgently. “Come, now!”

“I cannot,” she said firmly. “My duty holds me here.”

Dusk turned to chaos. Stria, watching prisoners huddled in rain, was confronted by Sophia’s sudden, defiant arrival.

Words sparred—faith against cruelty.

Sophia revealed a bowl of divine elixir, splashing it onto Stria, who shrieked as her necromantic skin steamed in agony and weakness.

“Apprehend her!” Stria commanded, but her guards faltered

against the light.

Sophia worked rapidly, undoing the corruption with the elixir, gently prying crystals from the now freed militia men.

Hope flickered amid shadow.

Chapter 19

Stria stood impatiently, waiting until her soldiers returned with Sophia, who was begging for mercy. She had waited long enough to regain her strength. She still waited, however, thinking they had followed her into some back room where she had hidden herself and was probably being dragged out at that very moment. It had only been fifteen minutes, but her patience was starting to run thin. She considered going inside, but knew she could probably expect Sophia with a cup filled with the divine elixir ready to splash it on her once more. This was, of course, unlikely but still possible.

Her mind then shifted to her clothing. If only she still had her cloak she could cover herself against any more of Sophia's attempts to splash her... Oh no! She thought as it dawned on her that her robes still lay inside on the bowl. Which, if Sophia removed...

Knowing she had sent her soldiers into a trap, she

immediately searched for a solution—any solution to get Sophia. First, she considered burning down the temple, but she had no materials to start a fire nor could she guarantee she could keep it under control. Marble and stone were not easy to burn anyway.

Covering her remaining men so they would not be affected by the light was her next idea, but again she lacked the necessary materials and manpower. Her prisoners might try to escape if the number of guardsmen dropped any further than it already had. So the only thing left to do was to go in herself and try to cover the divine elixir again so she and the militiamen still inside could focus on capturing Sophia. Now all she needed was a cloak.

She walked up to an elderly man with a long white-trimmed beard who was wearing a cloak to cover himself and his wife from the rain. She approached him and angrily demanded, "You there, give me your cloak now!"

The old man was frightened by her demands and started to

take off his robe hoping she would leave him and his wife further alone. He did not do it fast enough for her; she started to pull on the robe with the old man still trying to undo the knot he had made at the collar to keep it from falling off, almost choking on that same knot until Stria took out one of her sabers and with a quick motion cut the rope that the knot was made of, allowing the man to breathe properly again, with his wife crawling toward him to see if he was okay.

She then looked at the rest of the people cowering on the wet, stone-paved city street and said loudly, "Anyone else who wants to give me their robes?"

Several people started taking off their cloaks and giving them to one of Stria's guards, who was ordered by her to collect them, exposing themselves to the cold wind and rain on this depressing night.

Before Stria had the chance to put it on, however, the militiamen she had ordered to go inside the temple started to storm out. Now, however, they were no longer under the spell of their dark, purple necromantic crystals. Some of them carried small bowls of divine elixir. These three militiamen made a direct assault towards Stria, while the rest started to spar with their corrupted counterparts.

The first soldier threw the contents towards Stria who blocked it from splashing her by using the robe she had taken as a shield. The second militiaman threw the divine elixir towards her from her left, and the third almost at the same time from her right. She was able to cover herself from the left, but not the right, with the divine elixir splashing onto her back, which hurt and somewhat weakened her, but not significantly as her crystals remained dry.

She heard the unsheathing of a sword. The soldier in front of her stood ready to attack, as did his comrades who stood ready with their weapons as well. The soldier on her left carried a blade, the one on her right wielded a crossbow, ready to shoot her. She had no time to think, only a sense of instinct guiding her in that moment. She first threw the now soaking cloak at the man in front of her, landing on his face and stopping him mid-swing. The man on her right shot with his crossbow. Stria dodged it just in time. At the same time, the man on her left attempted to stab her with his sword, but she grabbed the sword with her metalized hands and kicked the man away with her powerful left leg. The would-be attacker landed several meters away, lying on the ground and gasping for air.

As she turned away from the gasping militiaman, she saw that prisoners were escaping in the commotion. For just a second she stood still, knowing that failure was not an option. The attacker that was in front of her came at her again, swinging from the side. She blocked the assault with the sword she had taken from his comrade. She blocked several more slashing attempts before drawing one of her sabers and slashing him. He fell to the ground immediately afterward.

She turned her attention to the soldier with the crossbow who was hurrying to reload it. Stria ran towards him. Just as he had loaded and was about to aim, he felt a growing sting in his abdomen just beneath his ribcage. He looked down and saw that Stria had put the sword of his fellow militia member into his torso. She pulled it out, and he took his last breath. She turned to the soldier she had kicked away, still holding his stomach and struggling to breathe. Seeing her approach, he crawled until he hit the temple's walls. He raised his left hand while his right still clutched his injury.

As she was about to give him the final blow with the sword that had taken his comrades' lives, she realized he was probably still of use to her. She knew she would lose some prisoners whether or not she ended this attack swiftly. So returning some of Drumir's lost sheep might help put him in a better mood.

As she cut down another of the now free guards, she made quick eye contact with Sophia, who watched the battle from a distance at the temple entrance, horrified as the men who had just been uncorrupted by the divine elixir began to feel the tide turning against them.

Staring back into Stria's emerald-green eyes—eyes like a predator stalking prey—Sophia felt her intentions and retreated into the temple, where the light of the divine elixir would protect her from her maniacal enemy. Though inexperienced in battle, she realized the tide had turned, and panic rose in her chest.

"RETREAT! RETREAT! EVERYONE RETREAT!" she shouted, intentionally keeping the temple doors open for shelter seekers, knowing Stria could not enter the elixir's light without injury, nor could her corrupted militiamen. However, Stria would not stand idly by. She intentionally attacked uncorrupted militiamen trying to retreat back into the temple—some she killed, others she wounded and captured.

When the last five guards made it back inside, they immediately closed the doors. Though the corrupted militiamen could not endure the divine elixir's light, the bolts of their crossbows certainly could.

The men looked to the door in distress, expecting Stria and her small cohort to smash through and slay them with overwhelming force. Some murmured holy scripture; others, less aware of the sanctity of the space, cursed.

Sophia, meanwhile, performed a religious rite at the bowl filled with the divine elixir. She heard curses even from afar but chose to forgive, given the dire circumstances.

"Gentlemen, they will not enter the temple, for the divine elixir's light shall keep them at bay," she said as she concluded the ritual.

"But they may try. That metallic woman got in here, did she not? All she had to do was put on a cloak and she was inside!" one soldier exclaimed.

"All the more reason to calm your minds and devise a plan," Sophia said calmly but firmly, making sure to look each of the remaining five in the eye.

She could see they were calmer but still anxious. Though not a general, she knew people were more capable if properly inspired.

"I helped a man escape through the temple's roof before that metallic woman and her corrupted guard came here—perhaps we can escape the same way," Sophia explained.

"When did this happen?" one guard asked.

"Just after that woman entered the temple and took the people inside hostage," she replied.

"I guess it's worth a shot," the young private said.

After barricading the temple doors with benches and two bookshelves—all books fallen out as the soldiers had tilted them forward to move quickly—the scene dismayed Sophia. She whispered quietly to herself, "Bunch of barbarians!" but held her peace.

The furniture was placed surprisingly quickly, though such small objects before massive doors was almost laughable. One bookshelf was already larger than the average person. Three stacked would be just shy of three-quarters of a meter below the ceiling. Still, the fact that anything barricaded the entrances, even ungarded, brought some assurance.

Sophia led the soldiers upstairs toward the basement escape hatch. The group made their way onto the temple roof. From there, they saw numerous corrupted guards patrolling the streets around the temple and entering nearby buildings, likely throughout the city.

They realized quickly, upon inspecting the temple's front—the site of the battle and retreat with heavy losses—that they had inflicted significant losses on the corrupt militia, around fifteen men—a notable share of the soldiers present despite their own losses. This count seemed to have grown with many patrols of three roaming the city.

"Where did she find all these?" Sophia asked, overwhelmed.

"My guess is that she pulled militiamen from guard duty at posts by the river and walls to patrol the streets in groups of three," explained a tall soldier with a dark-blond buzzcut.

"Well, I guess we won't be escaping after all," another

muttered.

"Well, could we not simply eliminate one patrol if they discover us?" Sophia asked. "I mean, we outnumber them in that case."

"We could, but by the time we finish one group, the rest would swarm us," the large soldier replied. "Besides, hiding bodies quietly would take too much time and effort we can't spare."

"So what are our options?" the younger soldier asked.

The large soldier, now clearly taking leadership, said, "We make a last stand."

"A last stand? Why? How?" Sophia asked.

He moved closer to the roof edge and said, "Look at that metallic woman and her large group of corrupted pawns. She wouldn't place them all there unless she's preparing to attack the temple. As for surrendering, given the tales of Drumir, I doubt we'd be treated with respect or honor as prisoners of war. As for how, we need to know our stockpiles exactly."

The five-person squad and Sophia retreated inside to plan their last defense. They chose to hold the upper temple levels; the narrow stairs made an excellent chokepoint, preventing Stria's troops from encircling them. The crossbowman took position by the supply entrance, though he had only about five bolts left.

"Why so few bolts?" the large soldier asked harshly.

"Mustn't have been resupplied when I was still one of those corrupted militiamen, I guess," was the reply.

"Wait... that means..." the large soldier trailed off.

"Is something wrong, sir?" the young one asked.

"No, no. On the contrary, this is good news," he said.

"How so?" Sophia asked.

"Well, think about it: if corrupted militiamen are sent out with fewer crossbow bolts than they can carry, it means the militia's weapons stockpile has been significantly depleted. The resources they use daily must be strained," the large soldier explained.

"Isn't it possible the corrupted soldiers just neglect maintenance?" Sophia wondered.

"They are careless, but they still understand how to use their weapons—even at a basic level. Getting new bolts shouldn't be too difficult unless they have almost none left."

"It won't make much difference now, though," the crossbowman said.

"But it's good news for our troops—which means there's still hope for Drune," the large soldier said. "Now, what else do we have in the arsenal?"

"Well, maybe this will help," Sophia said, opening a vat filled with divine elixir, its gentle light illuminating the room.

The large soldier laughed, "I know it will!"

Chapter 20

Arlen looks through the window while hiding behind a curtain so as not to be spotted by the corrupted guards patrolling the streets. As he watched, he saw the corrupted guards break into people's homes and drag anyone they found outside.

Arlen felt sweat break out on his forehead—it was only a matter of time before the militiamen would come to ransack his home and drag him off to who knows where. He knew he had to find a way out before they came for him. Fighting them head-on would be useless, for he was outnumbered and armed with only a knife, a prototype revolver, and a big bag on his back that was wearing him down with its sheer weight. The revolver shook heavily in his hand. He was hesitant to use it, for he had not tested it enough to be certain it would not blow up in his hand, as likely as hitting the enemy.

"Wait, that's it!" Arlen whispered to himself before racing to his basement. There, he opened a can filled with black powder. Carefully, he took some out and put it in a round-bottom flask, which he sealed shut with a piece of cloth propped inside, only a part of it hanging out. He tied some small bags filled with white powder firmly around the flask. Just as he was almost done with his homemade device, he heard his front door break open. Thunderous footsteps followed immediately. “Well, they are here,” Arlen mumbled. He looked at the basement door. He had closed it just in case they tried to break in, though he had no idea how long it would hold. Another reason to hurry and escape.

Before leaving, he took two fire-stones and struck them together close to the cloth hanging from the flask, which had been soaked in oil to make it easier to ignite. Sparks landed on the cloth and a fire began to burn.

The flame consumed the cloth quickly and it wouldn’t be long before it reached inside the flask.

With time seeming to slip away, Arlen quickly opened the basement gate, slipping back onto the streets just as the militiamen kicked it open.

He caught a brief glimpse of three militiamen racing toward him before slamming the gate shut and running for his life. Soon, he heard the pounding footsteps of his pursuers. He could not tell how many, but the loud stomping and clinging of their armor told him it was at least five. The more he listened, the louder they seemed to become.

Arlen ran as fast as he could, his chest heaving with heavy breaths. His bag thumped against his back in time with his footsteps, slowing him.

In his hands, he held his homemade inventions: the revolver in his left, the flask in his right. The burning cloth grew shorter; the flame's heat was hard to ignore against his hand. Letting go was no option—it was his only hope to escape the pursuers.

His heart pounded fiercely, but no matter how hard he ran, his lead was fading. Exhaustion whispered that it would soon overtake him. If only he could reach his boat and get off the coast, he would be safe.

Ahead, Arlen saw a road junction with a fish stall on the right. An idea struck him. He hurled the flask toward the junction, and it arced through the sky before crashing to the ground and shattering into shards.

With the glass broken, the flames engulfed the black powder, causing an explosion that shredded the bags of white powder tied around the glass. The shockwave spread dust across the near-empty street.

The thick mist that washed over the area was so dense that visibility was limited to a mere arm's length. Sound became the only sense; even smells faded except for the mist's bleak, musty odor.

This was what Arlen was counting on.

He hurried toward the stall, tearing it partially down to block the road with rotten fish before slipping into the house behind it, where the smoke and mist shrouded all in near darkness.

Though he had memorized the street layout, he knew nothing of the house's interior, which now felt like a maze hidden by the smoky veil. If it resembled his home, or at least its architectural style, he could guess the general location of key spaces.

After correctly guessing the staircase location, he moved carefully upstairs, making as little noise as possible to avoid alerting the corrupted militiamen still hunting him.

Each creak of wood sent a thrill of anxiety; he inched along until he found the window, identifiable by the faint light.

The window overlooked the intersection he had just fled; the militiamen were still in the street, spread apart and searching. One noticed the fallen stall and the others followed, disappearing around the corner.

After waiting a few more minutes to be sure they were far away, Arlen caught his breath and moved quietly through the streets.

Every slightest movement sounded like a crashing tree in his ears.

Slow but steady, he moved toward the harbor, frequently glancing behind, his nerves tormenting him. Had they seen him? Was it an ambush? Were more coming?

Sweat beaded on his brow.

After what felt like an age, he saw his boat. Slowly, he crept closer, keeping to cover.

The last few meters crossed a sandy beach, bare and exposed. Before he could sprint to the boat, footsteps came briskly from a street between him and the vessel. Arlen tried to stay as quiet and unnoticed as possible, inching closer before the guards could spot him.

There were still several meters and no cover. Waiting was impossible—they were heading straight for him.

His only option: run.

He dashed for the boat, pushing it into the water before they could catch him, preferably before noticing him.

When they did, for a moment, Arlen locked eyes with the

leading militiaman. Despite the man's fear and burden, Arlen saw a blankness at first, then a flicker of the trapped soul, desperate to end the madness.

Arlen threw his bag and revolver into the boat and pushed it with all his might toward deeper water. As the water reached his knees, he climbed in.

Before he could start paddling, a guard seized the back of the boat.

Arlen spun and struck the guard's face with the paddle, then his grasping hand. The guard did not surrender, swinging a sword, which Arlen barely dodged as it plunged into the wood of the tiny boat, fixing his grip.

With few options left, Arlen raised his revolver and shouted, "LET GO!"

Only vague murmurs and grunts met his cry. Arlen had no time to decode them—he focused solely on survival as the corrupted militiaman closed in. Two others marched through the water and mud, drawing closer and closer.

Desperation overwhelmed him; he pulled the trigger.

For a moment, time froze.

Arlen stared into the eyes of the man he had shot, seeing the man beneath the corruption slipping away before the body collapsed into the water.

There was no time for remorse; Arlen paddled hard, fleeing the last two pursuers with every ounce of strength.

Though he escaped, only one thought thundered in his mind: What have I done?

Chapter 21

Mynedrie still lay in her bed, thinking about what the commander had told her. As cryptic as it may have been, it seemed to have cleared a fog in her mind, at least a little bit. What does it mean? she thought to herself.

"Mynedrie, are you up?" a hushed voice asked. She recognized it—it was Arlen's.

"Yes, I'm up," she said, moving to sit upright.

Arlen placed a heavy bag on her bed beside her. As he opened it, he said, "Good, because I have a gift for you!" and took out the metallic arm.

"What is that?" Mynedrie asked.

Arlen held the metallic arm in his hands and said, "It is an arm that I made... for you..." As Arlen tried to carefully explain the arm's purpose, Mynedrie looked at it with amazement, whispering to herself, "The saying..."

"I'm sorry?" he said, confused.

"Yeah, I... I just remembered something the commander told

me," Mynedrie said with a sense of deep introspection. "How... are we going to do this exactly?"

"I am going to place it at your side and the arm will do the rest. I must warn you, however, that it may not work! But if it does, it is going to hurt—extremely," Arlen said.

"Well, as you may have noticed, I am no stranger to that," Mynedrie countered. She turned her shoulderless side towards him while grabbing her loose, unkempt hair with her right hand so it wouldn't get in the way.

Arlen moved closer with the metallic arm and placed it just in front of where it needed to go. "Are you ready?" he asked, concerned.

"Yes," she said, trying to sound as confident as possible.

"All right then. I need you to place this in your mouth," he said, giving her a strap of leather. She did as he said and braced for what was to come.

"All right, here... we... go," Arlen said before pressing the arm against her. Almost immediately after it touched her skin, a hissing sound and steam came off of her. The hissing was quickly joined by mumbled screeches of pain, so intense she fell back onto her bed, clenching her right fist and biting hard onto the leather strap.

Arlen quickly rushed to the other side of the bed, placing his hand into her clenching fist to relieve some of the stress. As he looked on sorrowfully, Mynedrie suddenly arched her back upwards, lifting it off the bed while even louder screeches left her mouth. Whatever the arm was doing now hurt her more than before, and all she could do was tough it out.

Arlen was right, however, as something different was happening. The hissing stopped briefly before small, glowing strings no bigger than sewing threads started spreading throughout her shoulder. She could think little, but instinctively feared the strings were burying themselves—her heart! her mind! In the end, however, they went to several different nerves and muscles hidden underneath the burned wood.

Then finally the agonizing pain faded. The relief was so great that with a huge sigh Mynedrie said in Moarla, “It's over, thank Miarha it's over.”

“How are you feeling?” Arlen asked, concerned.

“Besides the torture, absolutely fantastic,” she said, waving her left hand… She stopped mid-wave, realizing it was her new arm she was moving. With amazement, she looked at the moving machine, gently moving it around, opening and closing the fist while observing. It felt powerful yet off to her. Perhaps one of the measurements was off, or it was because she had gone without a left arm for more than a day, or simply the fact that the arm was not made of flesh and bone was making it feel odd.

“Mynedrie?” Arlen said to regain her attention.

“Hmm… yes?” Mynedrie responded, snapping out of her concentration.

“How does it feel? Your arm?” he asked.

"It feels... off. Not in a bad way. In fact, it feels strong, energetic, almost like it has its own heart," she explained. "It's difficult to put into words, but in a way, it feels almost alive. How did you make something so incredible?!" Mynedrie continued.

Arlen started to sit next to her and said, "That is going to be a long story."

Chapter 22

"So, to get this right, you went behind enemy lines to create a powerful energy crystal for a metallic arm meant for me?" Mynedrie said.

"That about sums it up, yeah," Arlen responded.

"Well, that was quite the adventure," Mynedrie said.

"Not finished, I'm afraid," Arlen said somberly, taking out his revolver and placing it between the two of them on the bed.

"Is this not your gun?" Mynedrie asked while picking it up with her new arm.

"Yes," he responded.

"So, what is wrong with it?" she asked.

"Nothing at all, it worked exactly as it should," he said, sounding almost let down by that fact.

"What's wrong?" she asked.

"I... it's nothing. Forget I said anything," Arlen said, looking away.

"It's not nothing, please tell me," she said, gently turning his face back to her.

"I... I killed a man tonight. One of the guards. He grabbed onto my boat, preventing me from getting away. I... I had no choice; he wouldn't let go!" Arlen said, his voice cracking through his stoic facade.

"Arlen..."

"I know how it sounds to you, but seeing that man's life leave his eyes... He had friends, family who will never speak to him again because of me."

Mynedrie could see Arlen was struggling with his conscience. "Arlen, this guilt you feel is understandable, but you have to realize that you had to!"

"I know that... But it changes little. I'm sorry I did not mean to place such problems on you—you have enough to deal with already."

An awkward silence fell between them. Mynedrie stared at the flickering candle on her nightstand and said, "I remember the first time I had to kill someone."

Arlen turned toward her, surprised yet interested in what she was about to say. "It was my father... A great and terrible plague swept through our hometown, turning people into an abhorrent mix of elf and fungus. All that was needed was a bite or a scratch from one of those things to be turned, and my father... was one of the unlucky ones to get infected... So I took him out of his misery." Mynedrie's eyes began to water as she said the last part. Arlen wrapped his arm around her to comfort her before she continued, "It's fine, I'm fine. I made my peace with it, and so will you, with time."

Arlen looked downward, not feeling confident that time would truly erase his guilt. "What also may help is meditation," Mynedrie said.

"All right, if you say so," he responded, sitting down cross-legged on the ground and closing his eyes.

Mynedrie moved around her bed to sit closer to Arlen, correcting his posture and lifting his face slightly. As she touched his chin with her right hand she saw a small smirk rise from the corners of his mouth. Staring at it, she heard a thought smolder in her mind, sh... should I?...

Before she could decide, they were interrupted by a knock on the door, followed by a question, "Excuse me ma'am, are you awake?"

Before she could answer, she was staring directly into Arlen's dark brown eyes, which gazed back into her soul. She averted her gaze and said, "Uhh... I, uh, yes please come in," trying to sound nonchalant.

A private entered and quickly looked at Arlen. "Ah, Mister Arlen, you're here as well. Good. I was looking for you both. The commander wishes you both to come to him."

"What is it about?" Arlen asked, trying to hide his worry that it was about his excursion behind enemy lines.

"He didn't say, sir," the soldier responded.

"We will be there as soon as possible. If I could get just a minute to dress, please," Mynedrie asked.

"Of course, ma'am," the private said, closing the door and waiting.

Mynedrie moved to her nightstand where several pieces of civilian clothing lay, given to her by the militia. Her own clothes were ruined—torn to shreds and covered by dried-up dirt. Whatever was considered fashionable in Drune, she did not know, but these clothes couldn't have been it. Then again, shortages were mounting, so there was no room to be picky. Letting men pick clothes an elven woman would find good-looking was an exercise in failure anyway.

She took out the best-looking pieces and placed them on her bed. As she was about to take off her nightgown, she noticed Arlen was still in the room packing his bag. Mynedrie let out a small cough to get his attention. They stared at each other for a moment before her gaze drifted to the door, which seemed to hint something to him, and he moved out.

Outside, Arlen and a militiaman waited for her to dress. An awkward silence grew between them, which Arlen tried to break, “So... was the militia your first career choice?”

“Oh no, I’m actually a tailor by trade. I joined the militia after the rats attacked the city,” he explained.

“Oh, interesting,” Arlen said, before silence returned.

After a moment, Mynedrie came out wearing a grey dress with a blue band of cloth dividing the silver-grey upper half from the brown-grey lower half. “So, how does it look?” she asked.

“Looks good on you,” Arlen responded.

“Honestly, not really your type of fabric,” the militiaman said with a disapproving face. “But no time for shopping—I’m afraid the commander has been waiting long enough,” he continued.

They made their way through the militia’s cold stone halls. Arlen wondered what the commander could possibly want. Was it about him? Would he face disciplinary action? He could only wait and hear.

Mynedrie, meanwhile, was less concerned about the

commander and more about the militiaman's snarky remark. What does he mean, 'not your type of fabric'? I mean, it's not a royal dress, but I'm making it work... right? Whatever—there are bigger things to worry about.

They reached the warroom where the commander and several of his lieutenants stood around a square table. A map of the city laid before them with several triangular pieces marked—a black saltire painted on the two upward rectangular sides. The triangles came in two colors: blue representing their soldiers and red representing Drumir's corrupted militia.

"Ah Arlen, Mynedrie, glad you could finally make it. Please take your places around the table so we can get started," the commander said.

As Mynedrie and Arlen sat, they noticed large chunks of Drumir's forces were placed in the city streets rather than at the river, where the front line—and likely the attack—would be.

Just as the commander began to explain, he noticed Mynedrie's new metal hand resting on the table along with the lieutenants who also noticed it.

Mynedrie looked at them, confused about what caught their attention, until she realized what they were staring at. "Arlen made it for me. Long story," Mynedrie explained.

"I'm sure it is, Arlen! Care to explain?" the commander demanded, clearly suspecting something.

"Well... I don't want to bore you with the technical details,"

"You won't!" the commander said sternly, as if looking right through Arlen's soul, searching for the truth.

Arlen straightened himself and looked back at the commander with a stoic face, knowing charm would help little here. "Well... I saw strange black strings running through Gritz's metallic implants. I concluded these 'strings' functioned like the nervous system, even integrating into it. I couldn't be sure they were safe for use in prosthetics, at least not for a non-corrupted individual. So I looked for a way to purify them to use in the arm you see before you. If I may brag, it's quite a feat of engineering and arguably even magic."

"Imagine civilians and veterans with damaged bodies able to walk again or shoot a bow like Mynedrie. Can you imagine the possibilities not just for us but for the entire world?" he ended with slight excitement, which seemed to infect the lieutenants and militiamen guarding the room. They looked at each other imagining the possibilities, except the commander, who asked, "Still doesn't explain how you made it," looking stern and folding his arms.

While internally panicked to hide his excursion behind enemy lines, Arlen answered, "I contacted an old acquaintance of mine."

"And where is this acquaintance now exactly?" the commander pressed.

"The temple," Arlen admitted, lowering his head in defeat.

"You're lucky we need you for this offensive, or I would

have gutted you like a fish for desertion! Understood?"

"Yes, commander."

"Now get out of my face!" the commander ordered.

Arlen left but not before smiling at Mynedrie, who couldn't help but blush.

The commander continued, "The western districts are in chaos—fires, looting, fighting. All of it. This works to our advantage as a large chunk of Drumir's forces are busy capturing people and putting out fires."

"What exactly started the fires and riots?" Mynedrie asked.

"We don't know, though my guess is Arlen had something to do with it. He's lucky he's a civilian; otherwise, I might have let him rot in the dungeon for years. Who knows, I still might," the commander said. "Anyway, this will give us ample opportunity. If we strike while the iron is hot, I've gathered our 500 most elite and experienced men to move through the sewers toward the southern and western city parts where most fires burn. Drumir's forces will be too spread and distracted to resist effectively."

"The second group will land from the sandbanks at the harbor. They'll face heavy resistance—the most difficult place to fight and the easiest to land."

"If all goes well, we'll flank them from two sides and force their retreat inside the castle walls. Any questions?"

"Just one, commander. It's a great plan, but why involve me? I have little military strategy experience," Mynedrie said.

"True, but you have more experience fighting Drumir than we do. So if this plan has flaws, I want you to find them," the commander responded.

Mynedrie looked at the battle plans. "I'd place some men outside the city's western or southern gates, just in case Drumir has a scheme up his sleeve. Your reinforcements could reach the rest of the group quickly."

The commander smiled. "Good—then we are one step closer to freeing our home!"

Chapter 23

Stria stood in the massive tent set up inside the castle's courtyard. She faced Drumir, who paced left and right, his hands clasped behind his hunched back. "So let me get this straight. You failed to round up some civilians and priests who had not only locked themselves in their temple but were also largely asleep when you arrived—somehow that turned into a full-scale riot, with fires raging all around our half of the city!" Drumir said, frustrated.

"Yes, sir, but many unknown factors were hampering the mission. Like how..." Stria stopped, looking into the yellow left eye of Drumir, his other eye hidden beneath his hood and his grey, crooked nose. His stare pierced almost directly into her soul.

He stepped closer and closer until his long-crooked nose nearly touched her face, hovering at eye level. His breath smelled of rotting corpses and decaying fungus—not from breathing, for he barely breathed.

Stria tried to keep her composure, but the more she looked into his eyes, the more brutal forms of 'discipline' seemed revealed to her. Her legs shook and her face was close to tears of despair. Every fibre of her body wanted to run, held back only by the fact that he literally controlled whether or not her heart would even beat.

"You are... lucky... lucky I do not have time to replace you right now! So I shall give you one more chance to serve me on the field and protect my current project," Drumir said calmly but intensely. "Understood?" he continued.

"Yes, sir," Stria said, hiding her inner fear as best she could with little success.

Drumir turned toward the giant sheet covering something big beneath it and said, "Good. Now leave!"

Stria obeyed immediately and left the tent without hesitation, heading straight to her sleeping quarters—which, by coincidence, were also the king and queen's quarters. She had taken it for herself, given that she was to become the city's queen after Drumir left.

When she first walked into the room, she felt like a majestic lioness. Now, however, she felt more like Drumir's working mule. The room felt wrong, almost mocking her very presence after how Drumir had spoken to her.

She went to the bed, and as she lay there, regret raced through her mind. She realized she had made a deal with an absolute monster, facing all the awful ways she could be 'disciplined' should she fail again.

She had fallen from the prestige of the elven nobility to a servant girl who foolishly turned herself into a puppet—with strings on her forehead, her chest, and even over her skin.

"Mama, Papa... please..." she whispered, knowing full well no one could hear her, nor would anyone if they did. For she had disappointed them too greatly to forgive or forget.

She recalled how they screamed, raged, and cried over her using that medallion. No matter what she said or how hard she tried to explain, they would not listen—instead continuing to scream in disappointment and shame. But above all, the utter disgust they seemed to feel for her hurt most deeply.

There was still a sprinkle of love in their hearts, giving her a small fortune before making her leave their home to somewhat save their reputation. It helped little for them, and made things worse for her. Any positive prospects in education, social standing, and marriage had vanished as soon as that medallion was revealed to the family. All thanks to her uncle Duran, whom she despised every moment of every day—along with his daughter and niece Mynedrie.

She could curse them all now. Not that it would help. Duran was dead, and so was Mynedrie, as far as she knew. All she truly had now were her feelings of frustration and tears running down her face.

Then she realized there was someone to talk to—not a friend, but someone at least respectable in society.

She wiped the tears from her metallic face and left her room, making her way toward the castle's dungeon, once filled with the city's worst criminals and traitors before Drumir took over. Now it was empty. All the prisoners had been sent upstairs and turned into loyal soldiers.

One cell, however, still held a prisoner recently placed after the raid on the temple—the only woman besides herself who had not been brainwashed by Drumir's crystals despite attempts.

There she sat in a corner, barely able to move. Her skin was goldish yellow with small cracks from which bright, warm yellow light shone out.

She raised her head slightly at the sound of her cell door opening, her eyes, whose irises had become golden yellow as well, staring toward the dark outline of the woman who had opened it.

"We need to talk!" the figure said.

Sophia struggled to sit more upright, using the dark grey stone wall behind her for support as she asked, "About what?"

"Well, you, to begin with. Why did you do it?"

"Do what?"

"You know what I mean—gulp—all that divine liquid down your throat?"

"Why do you want to know?"

"I will be asking the questions, understood?"

"Well, seeing how I don't have much time left, neither of us may get answers," Sophia said, pointing at her own face.

For a moment, Stria's left eyebrow twitched as anger rose then subsided just as quickly. She was all but officially the new queen of this city and yet was treated like the simple ship maiden she had been forced to become. "Fine, if you really need to know, I want to understand why... like this and not any other way?" Stria said.

"Any other way would still have allowed your master to use me as a toy for his whims," Sophia answered.

"That brings me to my second question: why stay at the temple? Surely there were better hiding spots?"

"Better than a building largely filled with holy light that hurts corrupted beings like you and those poor militiamen? I cannot begin to imagine how it must be for all those poor souls who live day and night with those... things on their foreheads making every choice for them!" Sophia said.

"And where did this need for freedom bring you? Right here! Alone in a dark, wet cell where you will spend the last moments of your life. If you had chosen to take the crystal, you only had to wait until I was officially made queen and you..."

"Hah! What a queen you would be. After your puppet master leaves the city (assuming he leaves anything for you) you would most likely keep using the crystals. If not, you would have put them back on the minute Drumir asked you to!" Sophia interrupted.

Stria smirked, keeping her head triumphantly high. "You know I'm a bit of a pragmatist. If you hadn't written your own death sentence, I would have given you a respectable position in my court—or at least as a respected advisor."

"If that's my only other road in life, then I gladly will spend my last moments here rather than serve at your side!" Sophia responded.

"I WILL BE QUEEN!" Stria shouted in a fit of frustration and rage.

"That doesn't matter. You can be a low servant, a queen, or even an empress, but you will always be a puppet on strings to him and nothing more!" Sophia retorted. "Honestly, you are just pathetic!" she continued.

"Is that so... then just you wait!" Stria said menacingly.

"Nice talking to you," Sophia said as Stria turned her back and left the cell. As she spoke, the stress from their talk formed a crack on her forehead through which light shone.

It was one side effect of drinking too much divine elixir. Other side effects included dry, sand-colored skin, which, while looking like fine sandpaper, felt surprisingly smooth and sparkled like glitter. Its elasticity would quickly dissipate, turning rigid and rock-solid, making movement harder and increasing the likelihood of cracks. Therefore, she moved as little as possible, trying to look presentable in case someone found her and recovered her remains.

She straightened her back, held her head high, and waited for the inevitable. Half an hour later, she heard loud footsteps approaching—it was Stria again. This time, however, she seemed to carry something peculiar: a crystal unlike anything Drumir had used before. It was as clear as new glass, with a golden thread running through it like a root through dirt, cut octagonally and about the size of her palm. Stria closed the door, and Sophia could see she was carrying other items as well—a strange metal hat of interlaced straps designed to be worn over the head.

Secondly, a strange medallion—a large circular gold-plated metal strap holding two small circular tubes between which the crystal would fit.

"Do you know what this is?!" Stria asked.

"A fashion statement?" Sophia said, snickering.

"You won't be laughing after I'm done with you!"

"What are you going to do? Give me a makeover?" she continued.

"You could say that I will indeed!"

"What do you mean?" Sophia asked, her glee shifting to mild confusion.

"Tell me, dear priestess, what exactly do you actually know about Drumir and necromancy?"

"I don't see what that has to do with this."

"Please, humor me."

"Well, seeing that I am a woman of faith, I do not meddle with raising the dead. But even the lowest commoner and simplest peasant have heard of Drumir's atrocities."

"Right, you know the tales about him, but there is so much you don't know. Like, did you know one of his hobbies? He likes to take people's souls and place them into special crystals that trap their consciousness inside. Like this one here!" Stria said.

Sophia's face shifted from shock to absolute terror. "Stria... no, you can't!" she pleaded, knowing what was planned next.

"That is the ironic part—despite being free from Drumir's bondage, I am now the one who will take control of your soul," Stria said.

"There is no one in control over my soul except myself and the divine!" Sophia said, her faith rising above her fear.

"Proselytize as much as you want. It won't save you!" Stria said, moving closer.

Sophia desperately tried to move away but could not; her body had already turned too much to stone to do so. All she could do as Stria placed the metallic hat onto her head, with the crystal trapped between the metal bars, was whisper to herself, "Faith has always protected me."

Chapter 24

Mynedrie was busy training with a militia bow taken from the militia's arsenal. As she shot all the arrows at the center of the target, Arlen came to talk to her. "Your aim is good."

"This is nothing. I used to hit that bullseye three times in a row at three times the distance!" Mynedrie said, frustrated.

"Well, if it's because of the arm, I'm sure we can..."

"IT'S NOT THE ARM!!!" Mynedrie interrupted. Arlen looked surprised at her sudden burst of anger.

Mynedrie's face meanwhile shifted from angry to ashamed.

"Sorry, I... I don't know what came over me," she said.

"Hey, it's alright. You can talk to me about it if you want," Arlen said.

Mynedrie did not answer, instead blankly staring at the table in front of her.

"Alright, but if you want to talk about it, you know where to find me," Arlen continued.

"It's my father!" Mynedrie said before Arlen could leave the training area. "He died when I was only sixteen years old, and it was my fault!" she continued.

Arlen kept quiet.

"I had to shoot him! Because of me, he changed into a pawn for one of Drumir's abominations named Thumir whom he defeated, even though he was wounded and was able to stop before it infected me as well. He saved me but I could not save him... and after all that I... I..." Mynedrie could hardly speak due to the trauma that the memories were bringing back up.

With tears running down her cheeks, Arlen placed her head on his right shoulder. "Shhh, shhh, it's alright."

"No, it's not. I'm a failure!"

"Would a failure have aided in liberating half of Drune and currently helping to save the other half?"

A long silence fell between them which was broken by Arlen who said, "Come with me."

"Why?" Myna asked.

"There is someone you should meet," he explained.

They made their way to the mess hall where some militiamen were having breakfast. One of the soldiers walked up to her and said, "Excuse me, miss. I don't mean to bother you, but I wanted to thank you for saving my life."

"I did?" she asked, confused.

"Yes. When we were hunting that ginormous rat, I got wounded and was certain it would have torn me to shreds if it were not for your intervention," the soldier explained. "My wife and daughter were so grateful that they wanted me to give you this," he continued, taking out of his bag a red scarf. It was made of a soft yet strong-looking material, and on both ends could be found a recreation of her family crest sewn with gold-like fiber.

Mynedrie took the scarf carefully from the soldier and said, "It's beautiful and so soft." As she rubbed over the fiber, it suddenly clicked in her mind that she was feeling it with her new metallic hand—specifically the palm and fingers made of pale-brown leather.

"Yes, yes it is, miss. It is made of steel silk. Royalty loves the stuff. Not only is it very soft, it's also very strong—almost as much as steel chainmail. The crests are made from golden wool, which is hard to come by but will boost one's luck if worn, or so they say," the soldier explained. "My wife and daughter were so grateful to you for saving me that they spent all night making it. Both me and many of my comrades put money together for it," he continued.

Mynedrie placed the scarf around her neck. It was thick and large enough to cover even her mouth with the crimson red fiber. As it covered her mouth, she could smell a combination of white daisy and gladiolus perfumes. The scent was strong but not overwhelming. She had developed a sensitive nose after years of searching with her sisters for flowers to decorate the house or make their own perfumes.

"Thank you, I love it. Thank you so much," she said, feeling that the words did not sound as genuine as she actually meant them.

"There is no need to thank us, ma'am. As a matter of fact, my cousins happened to take the road where we found that abomination an hour or two after you defeated it," one soldier told her.

"Or when you got us out of those tunnels," another remarked, quickly followed by a loud, cheerful rumbling from the rest of the militia throughout the cafeteria.

"Thank you everyone, I deeply appreciate it. As a matter of fact, tonight, first rounds are on me!"

This news made a loud cheer of excitement sweep across the cafeteria.

Mynedrie used this sudden outburst of excitement and the focus on drinking to sneak out of the cafeteria to find a place where she could be alone. She went into one of the toilets where she started crying with her right hand over her face while her left rubbed her new scarf. These were no tears of sadness, however, but rather of relief and gratitude that she had actually made a change for the better here even with everything that had happened.

Two knocks sounded against the bathroom door. It was Arlen, who asked Mynedrie, "Hey Myna? Are you alright?"

"...I... I'm alright," Mynedrie answered.

"Alright. If you decide to talk, I'll be in my lab with some food... if you want, you can come with me. Some food might do you good," Arlen responded.

Mynedrie said nothing, and Arlen simply walked back to the mess hall, ignoring the soldiers as he headed to his lab to work.

As he sat in front of his desk, he whispered to himself, "She is not alright."

Chapter 25

Stria looked from the balcony of the castle's light tower, from which she had a great overview of the city and the battlefield soon to come. She could imagine it now—how the harbor would be filled with the bodies of invading rebels and how her troops would drive any survivors back into the river. And after all her battling and suffering, she would finally become the Queen of this land.

There was still much work to be done, however. While the defense was nearly completed, she also had to stamp out the riots and extinguish the fires raging through the city, stretching her forces significantly. This was not good, not at all. She knew this was going to be a desperate assault—all the signs showed it. An important figurehead had died, their resources were dwindling, and with both morale too low, a coming attack was expected in an attempt to turn things back in their favor. If the military has shown anything, a desperate enemy is a dangerous enemy.

She turned her head toward the riots and looked on with rage, all while the fires' light illuminated the city. If only she had better handled her indiscriminate arrests, or not ordered them at all, she would not be here right now. Worrying would not help her now, sadly.

She made a mental note to use more ‘delicate’ ways as queen. That is, however small the change, she would lose, and Drumir felt the need to put whatever remains of her in some jar for the rest of her life, using her as a lab rat for experiments as he pleased.

She decided to move downstairs, the thought of what might happen if she lost motivating her to take close command of the defense herself.

While making her way down the stairs, she went over her defensive plans one more time. Her new heavy infantry took position at the front while archers behind them shot arrows in an arch over the infantry at the attackers, in relevant safety. Meanwhile, other archers and crossbowmen launched even more projectiles from above the rooftops.

If that line of defense happened to fail, they would still have to go through the second line, which was built with defensive works made by all the artisans and laborers she could scramble together as strong and as quickly as possible.

In case any enemy soldiers tried to attack from anywhere else along the river shore, she had several squads of eight soldiers each ready to give them a strong push back into the river.

Her defense was perfect, except for the fact that the riots and fires were still raging. If any rioters or fire came close to her defense, she could send a group of corrupted militiamen, numbering around twenty, to quickly deal with it and also act as quick reserves if her defense needed them.

As she walked outside the castle gates, she was met with the heat, light, and smoke of the fires raging all around the city, especially in the southern neighborhoods where the fires and riots were at their worst. Luckily, the riots and fires had been prevented from reaching the city square and the harbors.

On her walk from the castle to the second line of defense, she saw the construction workers, artisans, smiths, architects, and engineers either sitting or hanging around, tired from all the work. When they noticed Stria walking toward them, some of the men stood up.

The men all stood in two lines, one behind the other with an extra guy. It was obvious that the richer and better-clothed artisans, architects, and engineers stood in front, though even their fine clothes were covered in cement stains and sawdust. The men behind them were poorer, many clearly laborers, judging by their worn, stained, and damaged working clothes. Their body language further revealed their uneducated background—they stood with relatively poor and slouched postures, while the men in front still stood confidently. At least the nobles of this place understood the importance of posture, Stria thought to herself.

"It would seem you succeeded in your tasks, gentlemen!"

she said in a serious and harsh tone.

"We did indeed, ma'am," an old blond, graying-bearded man said.

Stria walked up to him. The man stood in the front row with good and proud posture. Despite his clothes being covered in cement stains, sawdust, dirt, and even a little torn, they were clearly expensive, and a pinky ring with a family weapon betrayed his noble background. "What is your name?" Stria asked sternly.

"Jacob of Astromir," he said with great pride.

"Well, Jacob of Astromir, how about you accompany me further in examining the defenses!" she said.

"But of course, ma'am," the man replied.

They both walked up wooden stairs, sturdy but clearly made of recycled wood from other objects. It was not the only thing made of confiscated materials.

At the top of the stairs, they found themselves on an improvised but sturdy defensive wall. "Well, ma'am, as you can see, this defensive barrier has been designed after the castle walls, with some liberties taken to make up for the limitations of available materials and lack of others.

"From here, your archers will rain arrows down on the oncoming enemies while staying safely behind their embrasures."

Stria walked to the closest embrasure near her. Looking out, she saw a downhill-sloping street paved with stone except right in front of the brick and wood-made wall, where wooden pikes had been placed at an angle against the street. Behind them were a few soldiers wearing armor made of thick felt and chainmail, carrying halberds—a powerful weapon when an average soldier tries to maneuver through wooden spikes and dodge arrows.

Stria and Jacob walked back down toward a thick, heavy barricaded door that led to the other side of the wall. Two soldiers raised and pulled away a big, heavy wooden beam resting atop two metallic hooks, which helped barricade the door against attackers.

Jacob opened the door, and Stria continued toward the first line of defense, which was at the market square, right in front of the piers. Of course, without any actual merchants to trade there, there was little trade to speak of. Even the stalls were taken away and used by builders in the defensive structures. The first line of defense was less impressive, but that was partly the point—to give the attackers a sense of despair seeing the big wall they had to conquer after struggling with the first line.

This defensive line was guarded by over twenty-one heavily armored, corrupted militiamen, recruited from the biggest and strongest men available. They were armed with halberds, making it a major challenge to come close to them.

Then there were archers and crossbowmen on the roofs, raining arrows down on enemies. If the attackers broke through the first line, there were plenty of archers hidden on rooftops surrounding the only accessible street leading straight to the newly constructed wall.

There was also a small wall—or rather a fence, a more accurate description—built right between the market square and the actual harbor. It was guarded by a few regular militiamen armed with a wild array of weapons: swords, shields, spears, halberds, bows, and crossbows. There were even a few maces thrown into the mix. They watched over the actual harbor filled with ransacked ships and boats, many with hulls cannibalized and cargo ransacked to make defenses. A small beach stood nearby, continuing underneath the dock and surrounding the wooden poles that held it up. Unbeknownst to the attackers, the sand hid a few unwelcome presents crafted by Drumir himself. He even showed some regret that he could not be there himself to "see the show unfold," as he put it.

Everything was ready. Stria took her place inside a building and moved to the second floor— a good vantage point to observe the battle while staying safe and able to give orders directly to the heavy militiamen standing in a three-line formation. She had also sharpened her sabers in case she needed to step in herself. She stood waiting for the inevitable attack.

Thoughts and doubts raced through her mind. Should she use this time to stop the fires and rioting? But what if they attacked while she wasn't here? Should she order the men guarding the fence to retreat and help the other militiamen? No, that would be foolish! She needed them here. The others would crush the riots, and she could rebuild the burning buildings once she was the queen. But what if they destroyed too much? Then she'd have to send whatever men she could spare to aid in firefighting after the battle. She could win this... she must win this... otherwise, instead of turning into a queen she would turn into scrap metal! If she was lucky...

She waited about an hour before seeing the first boats leave the northern side of the river and move toward her. Once in range, she ordered all front-line archers to shoot volleys of arrows toward the boats. The arrows flew in great arches before raining down like deadly rain on the oncoming militia. Most arrows hitting a militia member were deflected or stuck in their armor without causing injury. Some arrows pierced shields. Others were less fortunate, suffering serious wounds; some even fell off the boats and, weighed down by armor, sank to the river bottom, drowning as they went.

Despite drawing first blood, her adversaries bravely continued between the clanking and thumping of landing arrows. Not that it worried her—after all, they still had to face Drumir's "presents," the first of which was activated soon after some of the first militiamen made it ashore and stepped on a few of the file creatures that started to grow beneath the sand before crawling out. Not fully formed yet, beneath their wet, oily, dark skin were true monsters—completely hairless, scrawny to the bone, clutching essential muscles. The creature launched itself at the first militiaman it saw. First scratching at his chest, its knife-long claws struggled to pierce the armor. It then used its ginormous jaw to bite off his head, leaving only the headless body and his dented, partially pierced kettle helmet.

As it chewed, another militiaman stabbed it from behind. The injury released a black goo that sprayed out, but the creature showed no sign of pain. If anything, it grew furious and lunged at its attacker, pushing him to the ground. It kept the monster's massive jaw filled entirely with slightly crooked, sharp teeth from chomping off his head by pushing it away with his shield. Struggling for survival, the soldier saw the creature's small black eyes that looked almost like glass glare at him with such fury.

Luckily, the doglike creature was distracted by an object stabbing its right thigh—a spear. Despite the injury, the beast turned with such force that the soldier was pulled several steps to the left as he clutched the spear shaft, hanging on for his life. The beast kept trying to get closer until stabbed again by more soldiers. A small group formed around the beast, stabbing and slashing repeatedly until it finally died.

Similar scenes happened all along the beach and beneath the pier. Despite the dreadful casualties caused by the creatures, the brave militiamen pressed on.

While comrades fought horrific creatures amid falling arrows, the first group reached a small makeshift wall guarded by a small force of corrupted militiamen.

The top of the wall was covered with sharp objects—glass, scrap metal, nails, etc. Not that it helped much, as the concrete holding them had not fully hardened, and even if it had, their armor simply protected them.

The corrupted opponents' weapons were more successful in piercing and damaging armor. Stabbing with spears, swinging maces, and the occasional arrow were all used to hold back the oncoming men slowly climbing a ladder from beyond the shore.

At first, the wall seemed able to hold off the militiamen, but as their numbers swelled, pressure mounted on the defenders.

One corrupted militiaman fell, then another, and another, until enough militiamen got over the wall and surrounded the remaining corrupt soldiers, who then also fell.

But the battle was far from over, as they now faced the newly created heavy infantry.

Brought together by Stria and Drumir, they guarded the only road not blocked by heavy objects or rubble.

The heavy militia took position, holding their halberds points first toward the oncoming opponents. They were an imposing sight—their heavier, black armor crude but fearsome.

These guys should hold them off at least a little while... Stria thought, a self-congratulating smirk on her face.

The first soldiers to charge were quickly stabbed and taken down by the halberds. The heavy infantry's strength allowed them to stab through militia armor with relative ease, with only some militiamen able to block them with their shields, which nearly faltered under the repeated impacts.

Not wanting to lose lives unnecessarily, the militia fell back several meters, creating a shield wall while arrows and bolts flew at them, smashing into shields or deflecting off kettle helmets. Spears and halberds rose behind them like flagpoles, readying for a counterattack.

Despite heavy armor, the non-corrupted militia's halberds could get through the heavy infantry with relative ease. Taking them out was top priority to give the heavy guard a chance to last long enough against the wave of soldiers before the attackers could reach the second line of defense.

Stria ordered her archers and crossbowmen to focus on the halberds, but their arrows seemed to make little difference. Their armor, helmets, and occasionally lifted shields protected them from projectiles.

Seeing the group of halberds behind the shield wall steadily growing and already outnumbering her heavily armored halberdiers, she decided to take serious action herself. She jumped out the window, her impact shattering the stones of the road beneath her. She rose and unsheathed her sabers, dashing toward the enemies with great speed. Together with her increased strength, she broke through the one-man-thick shield wall.

Having pushed two men to the ground, she moved to her targets—the halberds still regrouping. Taking them by surprise, she began spilling soldiers' blood, ramming one saber into a young man's guts while using the other to block the shaft of a halberd. She then stabbed that militiaman with the very same sword she had used on his comrade.

Despite the initial shock, they began to group on her. Wanting to prevent capture, she turned and kicked a soldier using his shield to the ground. She ran from the growing crowd, bumping and pushing anyone in her way to the ground.

She reached a less crowded part of the plaza. Some militiamen had followed her, but most remained behind the shield wall preparing a counterattack or providing suppressive fire on archers and crossbowmen on roofs.

A militiaman armed with a bow tried to shoot Stria, only for his arrow to deflect off her armored skin. She quickly closed distance as he nocked another arrow, sensing she intended to kill him. Luckily, a comrade blocked her saber slashes with his shield, giving him time to reload, but a fellow archer arrived, shooting several armor-piercing arrows.

They both readied their bows to shoot Stria, who was at the point of killing their fallen comrade, shield broken and sword out of reach.

They fired bodkin-point arrows, which flew a short distance before striking Stria, who wore a light and expensive coat of arms embroidered with her family crest in white wire. Despite the short deadline, it was made within a night—she had instilled enough fear in the embroiderers to work without rest, though lack of sleep showed in imperfections, which she expected given human craftsmanship.

Her pants were less interesting: made from expensive

materials and tailored to fit. She wore no armor to maintain speed; her steel skin was armor enough.

Both arrows tore through the uniform fabric. One arrow deflected, but the other pierced further into her skin and lodged in her thigh, making her scream in pain. She looked at her left thigh, feeling her muscles rub over it, then looked on in shock, realizing more arrows would come. She quickly limped back to the heavy infantry line.

Seeing their enemy wounded, the attackers focused their fire on her; soon, two, four, eight, ten arrows struck her body, mostly her back but also legs and arms, slowing her further.

"Capture that woman!" she heard an officer scream, no doubt realizing they had a real chance.

The soldiers obeyed. One grabbed her right arm, but she threw him back. Another soldier with shield and sword tried to seize her from the other side; she swung her saber to hold him back.

Quickly, more soldiers mustered courage to try and hold her.

Just moments from capture, she desperately ordered, "Heavy guard, move forward and get me out of here!"

The brainwashed soldiers advanced to rescue her.

The shield wall moved forward to hold off the black-armored heavy infantry but were quickly cut down. They reached Stria, who was almost buried by soldiers trying to capture her. Though strong, it was hard to move.

Then she felt something warm and liquid run over her head—it was blood, and definitely not hers. She looked up and saw some of the heavy guard slaughtering the militiamen who buried her.

Seeing they could not dent the heavy guard, the rest of her would-be capturers retreated quickly. The heavy militiamen created a protective circle around her and slowly escorted her back to the defensive line, where no corrupted militiamen stood. Only corrupted archers and crossbowmen held off anyone trying to flank them.

As the heavy militiamen escorted her out, one of their

hallberd-carrying enemies used his weapon to attack, pushing it right between the improvised metal plates protecting his guts. The hallberd pierced his belly, dropping the corrupted militiaman to the ground. His scream of pain was muffled by his helmet before he died in a pool of his own blood.

His place was taken by two men beside him, making the circle smaller. Then another went down, making it smaller still, until they finally reached the defensive line, forming a two-rank-thick line, with the second line one soldier short.

Stria, wounded but healing from her failed preemptive strike, slowly made her way to the second defensive barrier. As she moved, her body slowly pushed out the arrows embedded in her.

Behind the defensive wall, she moved to where archers and crossbowmen stood ready to shoot anyone trying to breach it. As she waited for the oncoming enemy, she thought despite her setback, she could still hold their ground. Though her archers' effectiveness left much to be desired—they were nothing like elven archers, whose skill and equipment tended to be better than any of these humans—they would have to do. However, the heavy infantry seemed to hold their ground longer than she expected.

Just as hope returned, a corrupted militiaman with a huge cut in his face came to her and said slowly and almost soullessly, "Commander... the enemy... is attacking... from the south..."

"WHAT!?" Stria responded and immediately continued, "But how?!"

Chapter 26

The sewers stank of rot and despair, making the men cough and moan in frustration. They marched four men side by side through the filthy water, praying they would not catch a disease. Their marching boots were wet and dirty, their armor soaking wet, but their weapons were still clean as they held them above the dark, stench-rich waters.

Myna walked in front of them on the relatively shallow sidewalk, along with some lucky soldiers behind her who did not need to walk through the slop. She prepared herself mentally for the battle to come. No doubt the soldiers were doing the same or praying for good fortune. Whatever was going on in each soldier's head, each felt, at least partially, the weight of their coming actions on their shoulders.

She secretly wished Arlen were there. He always had something clever to say, but he was ordered by the militia commander to lead several small groups of inexperienced recruits, carrying his latest invention which was supposed to help put out the raging fires burning in the city's southwestern streets—fires that had turned wealthy and exquisite homes into smoldering husks.

The lieutenant next to her ordered his men to halt, which the soldiers obeyed. They stood right above one of the sewer pits leading to the streets above, many other groups doing the same throughout the southwestern sewer system.

The lieutenant turned to speak to his men. "Listen, men, these have been tough few days, but despite it all, we have persevered. But our task is not yet complete—oh no. Soon we will rise from these sewers, liberate our city from these invaders, and drive them out of our homes! Now let me ask you one thing: ARE YOU READY!!!"

"SIR, YES SIR!!!" the soldiers responded in unison.

Soon afterwards, the lieutenant led his men up the ladder onto the smoke- and debris-filled streets. Mynedrie stayed behind until all the militiamen of her group had moved up, which took quite a while. Not that she would spend that time twisting her thumbs; she used it to ease their nerves by yelling encouraging words, which the soldiers seemed to appreciate, responding with "Yes, ma'am!" and "You know it, sweetheart!" and similar attempts at flirting.

Once they had made it up, it became the turn of the militiamen carrying one of Arlen's new inventions: a hose with a big box on one end, with a crank on both sides to be turned by men. This pumped water into the box through a smaller hose, then through the big hose to the other end where it was pushed through a muzzle, creating a stream of water to help put out fires. The only regrettable part was that it used sewer water. If it stank now, Mynedrie feared what it would smell like after heating.

Coming up through the sewer, she saw just how much these water hoses were needed. Without hesitation, the militiamen began spraying the fires with stinking sewer water. The spraying only worsened the smell; Mynedrie could not help but vomit and felt her eyes grow watery.

Mynedrie moved toward the rest of the militiamen creating a defensive line on one of the main streets leading to the city square, blocked by a smaller unit of corrupted militiamen covering the retreat of the rest of their army into the castle. Right behind the militiamen, Arlen instructed some recruits on where and how to spray water. “Arlen!” Mynedrie shouted.

“Oh, Myna, there you are. I was wondering where you were,” Arlen responded.

“Yeah, there was a long line in the sewer which took forever,” Myna replied, making Arlen chuckle a little. “So, what’s the situation?” she continued.

“Well, from what I gathered, a small group of corrupted militiamen is blocking our path to the plaza, and our men are having trouble taking them out,” Arlen explained.

Mynedrie thought for a moment before answering. “We should try to get to higher ground, get a clearer view, maybe even place some archers on the rooftops.”

"Maybe if we use that building right there, we can get up on its roof," Arlen said.

"Good idea," Mynedrie agreed.

They tried to open the house's front door but found it too tough to get through, unsurprising as it was made of thick, dark oak wood. If Arlen was right, a heavy and expensive lock secured it—nothing short of a master locksmith could open it.

Arlen pointed to the window beside it. "There, break it!" he shouted over the noise.

Mynedrie punched the window with her metal arm, shattering it into many pieces while keeping her other arm in front of her face. The window was taller than her, with the sill a meter and a half above ground. Luckily, nothing fell on her as most of the shattered glass stuck in the frame.

She pushed away shards stuck in the lower window frame with her metallic arm. They both climbed through, the larger shards dangling above them, and prayed not to be hit by falling glass.

Once inside, they made their way up carpet-covered stairs to the uppermost floor, the staircase rectangular with a large open space between floors.

They reached the attic, where many high-quality pieces of furniture were stored—sofas, closets, chairs. Behind a closet was a window that let light into the attic.

They broke through the window to get to the roof above. After a small climb over tiles, they reached the flat roof.

From there, Mynedrie and Arlen could look over the entire plaza, seeing the castle's large gates open, many corrupted militiamen walking from all entrances into the plaza. The only places the corrupted militiamen were not abandoning posts were at the harbor (which Mynedrie and Arlen could not clearly see) and a small contingent defending the street beside them.

All this was organized and ordered by Stria, who stood at the castle gates trying to get as many forces behind it before closing.

Mynedrie looked at the line of corrupted militiamen, aimed her bow just behind them, preparing to shoot an explosive arrow. She focused carefully and released it. The arrow crashed down, immediately exploding. A shockwave pushed through enemy lines, pushing them aside, wounding and killing a few, causing their desperate defensive line to falter. The militia quickly moved to enter the city square.

Stria noticed this, nearly succumbing to the stress of seeing her potential fates and punishments inflicted by Drumir as it unfolded.

Just as she wanted to collapse, she looked up and saw a familiar face—Mynedrie—looking down on the carnage she had caused against her troops, who were down to just a few slain in battle.

As she looked up at her niece, who seemed somehow different, she couldn't help but feel glee at the sight—the possibility, the opportunity.

Without hesitation, she sprinted towards her niece, cutting down any soldier who stood in her way.

She rushed as fast as she could toward the building, entering by jumping through a front window, then rushing toward the stairs, trampling anyone in her way. She was full of envy, contempt, and anger toward her niece.

Meanwhile, Arlen and Mynedrie stood ready to fight Stria. The moment she showed herself on the roof, she would be attacked with full force—Arlen with his gun, Mynedrie with her bow—patiently waiting.

Suddenly, a large crack sounded, quickly followed by another, with part of the roof collapsing closer and closer.

They moved quickly to the brick and mortar front wall made specifically for support, standing on the stone-masoned top just before the last part of the roof collapsed.

They looked at the collapse and saw a figure rise from beneath—the roof breaker, Stria.

She looked gleefully at her niece, screaming, "I thought that rat had swallowed that arm of yours like a snack! No matter, I get to finish the job!"

"Too bad for that monster—I was too much for him to chew! As for you, it wouldn't be the first time you failed," Mynedrie replied.

They faced each other, waiting for the other to make the first move, staring dead in each other's eyes, waiting for the first to give in. The flames of their surroundings and each other perfectly reflected in their eyes. Neither blinked, not once, to avoid being caught off guard.

Suddenly, Stria threw her right saber, accompanied by an almost primal scream. Mynedrie barely dodged, seeing the sharp blade fly past and onto the city square, narrowly missing an unsuspecting soldier.

Turning back, she saw Stria now holding half a support beam, running toward the wall Mynedrie and Arlen stood on.

Mynedrie tried to stop her by shooting arrow after arrow, which ricocheted off Stria's steel skin. Arlen tried with his gun but struggled to load it.

The wall fell, followed by the floor Stria stood on collapsing.

Arlen and Mynedrie fell with it.

Arlen aimed his body onto a pile of rubble, breaking his fall, while Mynedrie timed her jump perfectly to land in the same hallway where Stria stood.

She immediately began firing arrows at her niece, who took them with no problem.

As Stria closed in, Mynedrie moved farther back, still firing arrows.

When pushed so far back that her elbow touched the door at the hallway's end, and Stria considered sprinting at her with full speed, Mynedrie shifted her aim to a section of the attic floor covered in debris. She focused on making a small explosive arrow, avoiding too big an explosion to prevent further building collapse.

With one arrow, she caused hundreds of kilos of wood, stone, and miscellaneous property to fall on her nemesis.

Mynedrie used the distraction to set her bow aside and draw her knives. Charging at her distracted niece, she slashed with her left blade, which Stria blocked with her saber. But Mynedrie's right blade stabbed unnoticed into Stria's left abdomen, piercing halfway through her metal skin and into her body.

Stria screamed in pain and retaliated by swinging her saber. Mynedrie ducked and quickly stabbed her in the right side. Stria screamed again, swinging her saber while Mynedrie crouched. Instead of dodging, Mynedrie caught the blade with her left metallic hand and held it firmly.

"It seems you have some improvements as well!" Stria laughed.

"More a necessity than an improvement, I'd say," Mynedrie replied.

"Sure, after all, you let your arm be gobbled up by a rat!" Stria taunted.

"And that rat came closer to defeating me than you ever did!"

Mynedrie countered.

Angered, Stria forced the saber to slide out of Mynedrie's metallic hand and kicked her back to the hallway door. To prevent the kick from breaking her ribs, she cushioned it with her metallic arm, which took a massive dent in the forearm. The kick pushed her so hard she broke through the door behind her.

Stria grabbed the knives stuck in her sides and threw them at her cousin. Mynedrie barely took cover behind a wall before the knives flew through the window. She felt trapped, unable to reach her bow or knives thrown out the window.

She was completely unarmed—literally—as her left metallic arm was so damaged she could only move her thumb.

Her right arm was fine, but the only weapons she could use were large wood splinters and shards of porcelain and glass on the ground—none of which could pierce Stria's metal skin.

Frantically looking for anything to help, she heard a loud

BANG followed by a painful scream, "AAAAAEEEEGGHH!!!" She turned to see Arlen pushed against the wall by one of Stria's hands, his other arm hanging limply. His saber lay next to a pool of Stria's blood dripping from a wound near her shoulder.

Arlen saw Mynedrie watching and quickly threw his gun to her, struggling to speak while choking.

"Myna... here!"

Mynedrie grabbed the gun and immediately aimed at Stria's forehead, where one powerstone was located. The fired bullet shattered the stone into shards that immediately fell from her head.

Disoriented, Stria raised her working arm to her forehead. As she removed it, several stone shards fell into her palm. She looked at Mynedrie, who still aimed at her, impressed by the weapon's power. Stria asked fearfully, "W-what did you do?!"

She did not answer, instead shooting again at the energy

stone in Stria's chest. Stria covered it with her hand, which left a hole, but the stone still functioned.

Stria immediately turned to retreat, followed by Mynedrie, who shot her in one leg to slow her down. Standing in front of an open space next to stairs partly ripped away by falling debris, Stria was grabbed by Mynedrie, forced to turn and face her.

Stria then shot twice with the gun at the powerstone, shattering it like the smaller one into pieces. Mynedrie continued, "Let me give you some help!" and pushed the nearly immobile Stria over the edge, making her fall several floors.

She looked down to see Stria lying on her back, barely alive, hardly able to breathe or move without the powerstones.

She noticed the walls in front of her start to turn pitch black, smoke rising—the fire had spread to this building.

She turned to Arlen, who was rubbing his throat. "Arlen, are you alright?" she asked.

"Yeah, I'm fine, but it seems we won't be for long," he answered, aiming at the walls near the stairs.

Mynedrie scanned her surroundings, hoping to find an escape route. She noticed big doors with large windows partially covered by thin, pale white curtains through which outside light shone into the chamber where she had been kicked.

"There! I think there's a balcony!" she said, pointing.

They walked to the balcony. Opening the doors, they were met with smoke-covered skies, buildings burned, looted, damaged, or all three, and a city square full of militiamen with bodies of corrupted counterparts scattered on the brick ground. The survivors had retreated into the castle.

Arlen placed his arms on the thick, stone-carved balcony rail, staring at the castle. "Almost. We are almost there."

Chapter 27

Arlen looked at the almost empty city square in front of him from behind the barricaded streets. The city square was filled mostly with the bodies of corrupted militiamen, but also a few unfortunate members of the militia itself and some of the large bolts that had been shot with the large ballistas on top of the castle's towers.

There was one on top of each tower and two faced towards the streets they had previously fought so hard to get through. These ballistas were strong—so strong, in fact, that Arlen purposely watched to make sure they did not directly aim towards him, for their bolts would most definitely fly through the thin layer of scrap wood and furniture placed just where the street and the plaza met.

"Care for some tea?" Mynedrie asked in a kind and still sleepy tone.

"Why yes, thank you," Arlen responded.

She handed him the cup of tea before asking, "Any

movement yet?"

"No, none whatsoever. I think they might be planning something," he responded.

"Yeah, the commander said something similar as well," she said, taking a sip from her cup. Despite being freshly boiled, it was not that good—even a relative simpleton in the kitchen like her younger brother could make something better. Still, it was better than no tea at all, even if barely so.

"Speaking of the commander, what is his plan exactly?" Arlen asked, taking a sip of his tea.

"Well, it basically boils down to three main ideas. One is that we try to starve them out until they either surrender—which I can tell you Drumir is never going to do. The second is to storm the walls directly, climbing on top of them with ladders. That one is expected to ensure the biggest casualties."

"And the third?" Arlen asked.

"The third was to ask you if you had any inventions in that beautiful brain of yours."

Arlen thought to himself for a moment before continuing, "Give me a moment to think of something, I may just figure something out." He then turned back to her and asked, "Speaking of, what do you plan on doing after all this is over?"

"After this... after this I will take a nice long vacation, which I will start by finding one of those bathtubs, fill it up with nice, steaming hot water and stay there for an entire day, doing absolutely nothing... you?"

"Me... after this I would probably do the exact same thing as you. A hot bath and a nice full plate of food," he said as he looked at her. "But first I would get a drink with you to celebrate..." he continued, his face at the end showing just a slight hint of nervousness.

"I would like that," she said with a warm smile, which brought one to Arlen's face as well.

"Excuse me, miss?" a young boyish voice asked her.

Mynedrie turned towards the voice and saw a young,

scrawny, and small boy standing next to her. He could not have been older than fourteen, too young to join the militia, so they let him serve as a messenger for the troops instead.

“Yes?” she asked him.

“I was sent here to tell you that we are done digging, ma'am, and to tell you that the commander wishes to see you there,” the young boy explained.

“Well then I will go immediately,” she told the boy. She turned towards Arlen and said, “I will see you later. And remember, the faster you get that idea, the faster we can go get that drink.” She smiled.

“Don’t worry, I will,” he said and smiled back.

She then followed the young lad to the building where she and Stria had fought their last duel. It was now a charred ruin with rubble lying all around. Much of it was badly charred, as was the building itself. Inside the building there was a large pile of rubble that was made up largely of the now collapsed staircase, beneath all of which was Stria.

The last time she saw Stria, she was dying, surrounded by rubble and debris that she helped create. Now she was a corpse hidden between the rubble of the now ruined home that had burned down and engulfed her body when it collapsed.

She had asked the commander if he could spare some men to dig it out to ensure that Arlen could 'study' it. Actually, she just wanted to prevent it from being defiled by an angry mob after all this was done. She was a traitor and a lackey of Drumir, but she was still family, and this was therefore more about the dignity of her family—and by extension, her own—than Stria herself.

As she got closer she noticed that the commander was not in a good mood. Something was bothering him, and it was obvious that Stria's corpse had something to do with it. He turned his gaze and then said in a harsh tone, "Mynedrie! Come over here!"

She placed her normal hand on the string of her bow that ran over her torso, while the bow's arms and handle ran over her back, rubbing the string with her thumb while holding it firmly in her hand, and asked the commander, "What's wrong?"

"You tell me!" he snapped back at her while aiming his arm to where her niece's body should lay.

She made her way towards the commander, and as she moved closer she could see that her niece's body was gone. The only thing that betrayed that she was ever there were some shards of the powerstones that she had broken in their fight last night.

"Wh... what happened?! Where is she?!" she asked, with fear and paranoia shaking her voice.

"Again, you tell me. My men have been digging and moving debris all morning to find nothing!" the commander said.

"What? You don't honestly think I had something to do with this?!" she asked, feeling insulted at the suggestion.

"Of course not, but that does not change the fact that an

enemy combatant's body has mysteriously disappeared. That means that she has gotten away and is trying to get back to her troops inside the castle or try to sabotage us all by herself, or worst yet, they have a secret entrance and exit through which they were able to reappropriate her body, and with it are able to send troops on a surprise attack, and we will be able to do almost nothing about it!" the commander explained, his frustration taking control for a moment.

"But you were the city militia's commander, should you not know about such secret entryways?" Mynedrie asked.

"Even if I knew every secret entryway and hidden stash of valuables that those noble men and women hid in that castle, what makes you think that Drumir's character does not simply create a new one with the help of his necromancy?!" the commander argued back.

"He is powerful, yes, but even he has his limits," Mynedrie explained.

The commander turned his back to her and scratched his chin.

He then turned back again and said to her, "Tell your boyfriend he has till tomorrow morning to get me something, otherwise I have no other choice but to attack with full force! Now go!" He then waved her away.

She moved away from his presence, her head now filled with many concerns: her niece's body had disappeared, a battle was about to start that she assumed from the way the commander acted stood little chance of winning, and did he just call Arlen her boyfriend?!

As she walked to the harbor where several boats stood ready to take people to the militia headquarters or otherwise the northern district, she passed by many militiamen, many of whom looked at her with respect and almost reverence. It made her a little uncomfortable, but also kind of good—after all, she'd been through much in this town, and she rightfully earned some respect.

While it was Arlen's lab that was most likely her final destination for today, it was not the only one, as she first had to make a stop at his aunt's tavern. By some miracle, she had been able to evade capture by the corrupted militia's attempts at capturing as many civilians as possible. Not a small feat for an old and large-sized lady like her.

As a matter of fact, she had helped some people hiding from them as well and allowed her tavern to be used as a resting place for free! The drinks had coincidentally gone up in price, however.

As she walked into the tavern, she saw Aunt (name) inspect his throat, no doubt inspecting the bruises that Stria had left behind.

"Please Aunt Anna, I'm fine, truly," Arlen exclaimed.

"Well, your neck looks pretty bruised to me! But I suppose it is nothing that will not heal on its own," Aunt Anna told him.

"Myna!" he said as she walked through the door. He walked closer to her and started speaking again, "So, how was it?"

"Not good, not good at all. Stria's body—it wasn't there!

Now the commander has told me that you have till tomorrow morning to come up with something, or he will order a direct assault on the castle itself!" she explained.

"But that's practically suicide with all the men they have in there!" Arlen reacted.

"I know!" Mynedrie said.

Arlen took a seat next to her and started to close his eyes while letting a hand run over his face. He thought and thought until Mynedrie whispered his name in worry, "Arlen?..."

"I... I'm fine... but how am I supposed to create such a game changer in less than a day? Even developing something like that often takes more than a day!" he said with frustration.

"Too bad we can't make more of these things in such a short time," Mynedrie said, while taking out the gun Arlen had given her.

Arlen looked at the gun and let what Myna said brew for a moment in his head before coming to a realization and saying, “Myna, you are a genius!” He then quickly got out of his chair and moved to his aunt and said to her, “I gotta go Aunt Anna, have a good day,” before hugging her goodbye.

“That’s alright dear, success with your work,” Aunt Anna said.

“Come with me, we’ve got work to do!” he told Mynedrie as he walked toward the door.

“You might want to tell me what you are going to do?” she asked, confused.

“I’ve got no time to explain; first things first, I will need you to go to the commander and tell him that I will need all bronze, cast iron, steel, every caster, every smith, and even a few stonemasons to come to the training ground inside the militia headquarters!” he told her. “Oh, and some stone pillars and all the tools that the men will require. I will send a note with the exact length and size of the pillar later,” he added.

Later in the day, Arlen was busy commanding the artisans

under his command left and right, making sure that no mishaps happened, as even the smallest misstep could make it all have been for nothing.

The stonemasons were busy turning pieces of rock into smooth pillars with rounded tips. Arlen could have sworn they made childish jokes about it when he wasn't around.

The smiths were meanwhile busy making a metal tube of steel where the stone pillar would be placed in, right inside the middle before they poured molten bronze inside of it.

That was the job of the pourers, who mostly dealt with iron ore from the iron mine. Still, they diligently inspected all the bronze that had been collected for impurities and removed them before melting began.

All of them labored for hours with little to no rest, knowing what was at stake. As the sun rose and fell, they continued their labor. A few other men joined them in the meantime as well; they were woodworkers tasked with making a frame to carry and transport the devices with relative ease.

Once they finished their tasks, the casters and the stonemasons were put to work making balls the size of a large apple, making as many as possible with their respective materials. The stonemasons unsurprisingly with stone, and the casters with cast iron.

When the smiths were done with their metal tube, they placed it exactly where it needed to go.

When everything was secured and put in place, it was the pourers' turn to get into action. They melted the bronze and warmed it even more at the militia's smiting forges. It was allowed to rise slightly above boiling to ensure it would pour properly.

They had to work quickly because the container had been filled to the brim. It was usually used to melt broken swords and spear points to forge new ones.

They first lifted the container by attaching gigantic hooks on each side, which took the strength of three grown men, then poured metal into the makeshift cast. Usually, the steel would be poured into a relatively flat mold to be easily divided into many steel bars.

They poured the molten bronze inside the mold being guided by three men—two experts and Arlen, who held a long stick with a metal sheet to ensure even distribution, as too much pouring at once could cause spillage.

Slowly and carefully, they tilted the container, letting a stream of glowing molten metal flow down into the mold, filling it slowly but surely with hot bronze. Arlen kept his gaze on the hot liquid metal in front of him, knowing the slightest splash could horribly wound him.

As he focused on the stream, Mynedrie watched the metal cylinder that also functioned as the bronze's mold, praying under her breath that it would not melt like the bronze. Arlen had told her that steel could take heat bronze could not, but it did not ease her nerves.

All the men working on it looked with great interest as well, waiting to see if they could reap the fruits of their labor. They waited until the very last bronze had poured before Arlen let out a huge sigh of relief.

Arlen then looked at Mynedrie, who looked almost as relieved as he did, and walked up to her, saying, "Well, that's finally over."

"So, what do we do now?" Mynedrie asked.

He looked at her and said, "Now? Now we wait."

Chapter 28

It was early in the morning. Mynedrie was silently waiting with a large group of archers for the signal to attack the castle. They were hidden atop one of the buildings, carefully remaining out of sight until the time to reveal themselves came. While the buildings were not as tall as the castle's towers, they provided enough elevation to properly aim at the militiamen controlling the ballistas on top.

Meanwhile, Arlen stood ready with a small group of men who had helped construct his new 'cannon,' as he called it. It was not a pretty thing, nor was he certain if his calculations held up, but he knew he had to aim it toward one of the ballistas. Calculating angle and distance, the cannon stood ready to blast the tower out of existence.

The plan was to first let the archers shoot at the walls and the ballistas, covering the cannon so it could load and aim at the ballista and turn it into splinters. They would then repeat this against the second ballista as well.

They waited for the sign to strike the castle and liberate it from Drumir and his corrupted militiamen once and for all. What felt like hours passed until the commander gave the go. When it came, the bow- and crossbow-armed soldiers unleashed a massive volley that took their enemies on the castle walls and towers by complete and utter surprise.

Quickly, the corrupted militia's bowmen and crossbowmen on walls and towers fired back, as did the powerful ballistas, whose bolts flew with such force they pierced the walls of surrounding homes.

However, a new contender had entered the battlefield: Arlen's cannon, armed and ready to shoot at the ballista on the right tower, which was focused on enemies hidden inside homes surrounding the castle.

The battlements atop the castle's tower hampered a direct trajectory to the ballista, but according to Arlen's calculations, the 'cannonball' could be shot with such force that it would smash through them and destroy the ballista.

To launch the cannonball, they had made a small hole in the back of the cannon and placed a fuse in it. The fuse quickly burned and ignited the black powder inside.

With a loud burst of air and sound, the cannonball shot toward the top of the tower, missing by only a meter from the intended spot. The men quickly adjusted the screw jack attached to the cannon's front while others frantically loaded the cannon as the ballista began turning their way.

The ballista, now armed and aimed at the cannon crew, shot its bolt directly toward them, nearly hitting the young man loading the next cannonball. Arlen barely pulled him away, saving him from being impaled by the spear-sized bolt.

They quickly lit the fuse again. This time, the cannon crew stood as far away as reasonably possible and covered their ears with their hands.

Again, a cannonball flew through the air and hit the target, flying a little past it, slightly misshapen from impacts. It landed just next to the massive lighthouse, flying just over

Drumir's tent.

Drumir had been working on his project inside the tent, not really bothered by the events outside, but the loud bangs and subsequent sound of shattering stone and mortar piqued his curiosity, betraying that something powerful had been used. He did not want to waste time on searching for the source and ordered a guard to investigate beyond the walls while he continued working on his massive project nearing completion. Soon, those beyond the castle walls would see what true power looked like—if only he could delay their advance a little longer, which wouldn't be a problem as long as his surprise awaited inside the castle's keep.

The guard returned with an arrow in his side, exclaiming, "Mighty weapon... left of us..." before falling unconscious, blood pooling beneath him.

"A weapon?" Drumir muttered. Whatever the weapon was, it was clearly a threat to him and his creation while unfinished. He left the tent and walked to the walls where fighting was heaviest.

As he climbed the stairs, he heard another loud bang, this time aimed at the other ballista, which it barely missed, causing massive stone chunks to fall just outside the castle.

At the wall top, Drumir tried to look at the device causing chaos, but his view was blocked by a shield wall protecting the cannon crew, as they reloaded.

He stood waiting for the cannon to fire again. The ballista shot back but hit the shielding wall, the bolt digging deep but not injuring anyone.

The men quickly reloaded and fired again; this time the cannonball hit and severely damaged the wooden ballista, sending shards flying.

Meanwhile, arrows and bolts began to rain down on Drumir. None hit him as they were blocked by a swirling forcefield of dark green light. Wood shafts touching the shield quickly rotted and decomposed, thanks to the talisman on his person. It blocked any attack closer than a couple centimeters, and anything organic would instantly rot on contact.

They tried repeatedly, but nothing hurt him. He smirked widely, happy to see his enemies in despair over his perceived invincibility.

Having seen enough, he returned to his tent to continue work, though more remained to be done.

Descending the castle walls, he was met by a cannonball that had gone through the battlements, killing two soldiers. Whatever the device was, he wanted to figure it out. He thought this was as good a time as any and ordered the surprise sent to the bailey, gleeful about what awaited enemy soldiers.

Meanwhile, Arlen was reloading his cannon again. He'd started getting the hang of it and sped up the reloading process. As men loading the ball called to him, "Arlen! You should take a look at this!"

Arlen moved to the cannon's front and saw a large crack had formed in the bronze cannister. He looked worried before turning to a young man and saying, "We'll give it one more shot and decide from there."

"Yes, sir," was the reply.

Having reloaded, he attached a burning rope to a long piece of wood and ordered all men to take cover behind a corner nearby. He lowered the rope to the ignition point, making contact a second later. At first, silence—then steel and bronze screamed in agony as the cannon ripped open in a massive explosion, sending shards flying through the air and burying in the surrounding walls.

Arlen checked quickly if he'd been hit. Luckily, he was behind the building a few seconds before the explosion.

Myna rushed to the side, yelling, "Arlen?! Are you alright?!"

"Yeah, I'm fine. Nothing to worry about."

"Anyone else hurt?"

"Just my stick," Arlen said, pointing to a stick lying on the ground.

"What... never mind. I suggest you get back to the militia.

Now that the ballistas are gone, the commander will most likely send in ladders and the battering ram," Mynedrie said.

"Good idea, we will," Arlen replied.

Soon, two groups of soldiers entered the plaza—one carrying ladders slightly taller than the walls, the other pushing a battering ram with a makeshift roof to protect against arrows and boiling oil thrown from above.

The battering ram stood just in front of the first gate, ready to ram it when it suddenly opened. Soldiers witnessing what was behind stood frozen in shock; before the rest could look, they were charged by a horde of screeching, ferocious wild people wearing the clothes of men, women, and children. They quickly overwhelmed the soldiers operating the ram, biting and scratching without mercy.

They ran and climbed over the battering ram and each other to reach all militiamen in the plaza, who had started climbing ladders. Soldiers on the ground retreated to the streets they entered the plaza from, while those on ladders soon found themselves surrounded by the crazed mob until their ladders fell or were pushed away by corrupted militiamen above. Some men on ladders tried to hang on battlements, but soon fell to the ravenous horde, which tore them apart.

Mynedrie watched in horror. Without much thought, she left her hiding spot, loaded an explosive light arrow, and shot it at the gate where the horde poured out. She caused many casualties, but the horde kept pouring despite it being one of the biggest explosions she could get from her bow, leaving her exhausted momentarily. She would have saved it for Drumir himself, but if the horde was not stopped soon, they would lose the entire western half, maybe all, of the city.

Arlen saw the militia forming a defensive line. Experienced soldiers quickly took position, but inexperienced recruits—who had nearly no training—were confused in the commotion, causing the line to be incomplete, with only the front facing line fully ready.

When the commander learned of the counterattack, he

immediately ordered all troops into defensive positions.

The troops held for now but soon began taking casualties.

The commander ordered a slow retreat deeper into the streets to ensure corrupted militia could not fire on soldiers on the ground or archers on roofs, who then rained arrows on the horde below without fear of return fire.

This did not fully remove danger, as building entrances faced the plaza and windows allowed the horde to break in and move toward archers and crossbowmen. The archers heard their awful screeching grow louder as the horde breached defenses. Lower-level soldiers met the terrible fate of previous comrades; some tried barricading rooms or floors but were overwhelmed, taken down by hands, nails, mouths, teeth, and rage.

Men on roofs fared better; the horde was slowed mostly by the limited entrances near roof edges where soldiers stood firm, shooting raging civilians clawing to tear them apart.

With the battlefield a battle of attrition, the commander and lieutenants brainstormed. Their options boiled down to retreat or stand ground. Neither was attractive.

Then the commander realized something and ordered Arlen to him. Arlen complied and listened.

"Arlen, you often make weird devices at home, right?"

Arlen wanted to correct him about 'weirdness' but bit his tongue and said, "Yes, sir."

"Could anything of those help us now?"

"Maybe, sir. I'll look."

"Very well. I'm assigning ten men with you to retrieve anything useful."

After turning, the commander added, "And Arlen, hurry! These men are dying there!"

"Yes, sir!" Arlen said and left quickly.

He and the men, some pulling a small wooden cart, made for his house in reasonable shape. The front door was broken, but the fires had luckily not reached it.

Arlen and militiamen entered the house, except four who

stayed with the cart. Arlen immediately went to the basement seeking anything to help frontline soldiers. He requested the designs from his study upstairs. He looked through everything, hoping to find anything small overlooked or deemed unimportant.

He first checked drawers and closets but found only materials. Eventually, he looked behind, underneath, and on top of them, finding only dust.

Finally, the soldiers sent to get his designs returned and quickly looked for anything worthwhile—the early draft of his 'pistol,' which only he and expert craftsmen could make from the designs; bullet plans, worth little without the pistol; the water hose and pump design, but spraying water on the raging horde would do little; then plans for the cannon, looking better and likely more durable.

Then he had a sudden realization: the water pumps could be used to pump something else—the divine liquid inside the temple. It had weakened Stria; perhaps if used against corrupted militia and the mad civilian horde, it could drastically change the tide.

He immediately ordered troops to move back to the commander's temporary command center, except for five who were to bring as many barrels of divine elixir as possible.

Arlen rushed to the command center to speak with the commander and lieutenants about his plan.

"Using divine elixir as a weapon and displacing it with hoses!? Do you want to anger the divine?" one lieutenant protested.

"Well, sounds better than letting Drumir defile it whenever he wants," Arlen argued.

The debate ended quickly, though unease remained among officers and Arlen, who knew it was either this or letting Drumir reclaim what they fought so hard for. They were not disrespecting their faith if protecting the innocent.

Arlen ordered all vats of divine elixir placed side by side ready for pumping. He also ordered the same five men to bring more barrels quickly.

The water hoses were readied; men assigned to spray stood three men behind the front line atop boxes to better aim. Two others pumped machines diligently knowing their and their comrades' lives were at stake.

Divine elixir began pumping from barrels. When empty, another militiaman placed the hose in a new barrel. At first, nothing happened; then the elixir burst out in fury, flying over men in an arc of pale divine retribution.

Zombies hit began to degrade immediately into corpses, falling as little more than piles of bones and flapping flesh. Corrupted militia resisted more due to armor but were eventually soaked through cracks or clothes, turning into sponges filled with divine elixir, falling in anguish as crystals' corruption desperately resisted purification.

This formed holes in corrupted militia defenses which the militia exploited. Quickly, the defensive line in the street where Arlen armed his cannon began splintering into groups surrounded by militia. Soon, other streets broke as well.

When Drumir heard this, he raged and threw things, furious they were losing after almost breaking the enemy. He ordered men back into the castle but let zombies continue stalking inside.

Hastily and disorganized, militiamen retreated. In the chaos, corrupted militia trampled their own trying to enter.

The commander could have ordered an attack on the gate but decided against it, knowing losses would be too high.

He ordered guards armed with swords and shields to round up zombies inside buildings, weapons best suited for space constraints.

Their help came just in time; archers began running out of arrows, crossbowmen out of bolts, some drew shortswords preparing a last stand.

Then came sounds of battle—blades slashing through enemies with such ferocity no zombies tried to reach them.

The fighting stopped briefly, then a militia member emerged

from a hole where undead had spawned, greeted by archers and crossbowmen ready to shoot—or men with shortswords standing ready.

Their nerves were bombarded by screeches that seemed to drive some insane—until it stopped.

Still nervous, guards held bows and crossbows ready.

"Don't shoot! We are your allies!" a voice yelled, and militiamen lowered weapons but stayed alert.

Several militiamen emerged, followed by Arlen, whom Mynedrie gladly welcomed with a hug.

"Arlen!" she said, happy and relieved, before embracing him back with a small smile.

Chapter 29

Mynedrie looked on with grief at the hundred-plus blankets covering the bodies of fallen soldiers. Widows, mothers, fathers, brothers, sisters, sons, and daughters were mourning their deceased loved ones or still searching for their bodies, desperately hoping their brother, son, or father was not among the dead. Among the small crowd were priests and priestesses performing their duty in preparing the bodies for funerals in the Skull Forest.

As she looked down the densely packed street, she rubbed the scarf given to her by a kind soldier. She wondered if he lay among the dead—the question nearly driving her mad.

She began to move slowly through the long, corpses-filled street, which seemed endless. Most bodies were hidden beneath blankets, but some had notes attached—people trying to identify the corpses of their lost loved ones. She thought she might find the soldier this way if he were no longer alive, as surely his family would be called on to retrieve and identify his body.

She also wanted to return the scarf he had given her; keeping it after his death felt unjust, as the only reason she had received it was for saving him.

She walked on slowly until she saw a woman and two children crying by one body that could be the soldier she sought. Approaching quietly, she first got a good look at the man's face before deciding to speak. To her shock and dismay, it was indeed him. His lifeless body was surrounded by loved ones crying streams of tears.

She knew it was not the right time to return the scarf, but keeping it did not feel right either. She tried to move on with as much dignity as she could from the mourning family until suddenly she was called out.

"Excuse me, miss?" It was the widowed wife of the fallen soldier who stood, wiping tears from her face. "You're the elf woman he told me had saved his life and made that scarf for?"

"Yes ma'am, that is correct," Mynedrie replied. Though the woman's body language did not suggest blame, Mynedrie still prepared herself.

"I want to thank you for all you have done," the lady said.

"Thank me? Why?" Mynedrie asked.

"For helping us free our country. My husband told me all about you."

"Yes, your husband. He was the one who gave me this scarf."

"Do you like it?"

"Oh yes, I can tell it was made by master tailors."

"Oh, you're too kind, miss, but I am glad you like it."

"I must confess something, however."

"And what would that be, miss?"

"I actually came here to return it to you because after your husband's passing, it doesn't feel right to keep a gift. He only gave it for saving his life in the first place," Mynedrie explained, taking off the scarf and trying to give it back.

The new widow gently closed Mynedrie's hands around it

and said, "Please keep it. It's not just a gift for saving one life but for giving countless others hope they thought would never return."

Mynedrie was speechless, unsure what to say.

"You're clearly tired, dear. Go get some rest," the woman said before returning to her children.

And so Mynedrie did, returning to the militia headquarters. As she entered the halls, she was reminded of when she was in the kitchen trying to cut fruit without her left arm—quickly reminding her of when Gritz bit it off, eating it like a cut of meat. Then memories and thoughts crashed over her: the dead battlefield, the vermin-infested sewers, the pain of a ripped-off limb—all melded into a cocktail of pain and regret.

As she moved through the halls, tears welled in her eyes. Not wanting to be seen crying, she started running down the hall, covering her face with her arm. Soon memories of her father's murder and the greyscale infection of her village flooded in, increasing her tears.

She slammed her room door open and then shut it behind her. She curled up in a ball beside the door and began crying relentlessly, feeling utterly alone and a mess.

She must have cried for hours until a knock came on the door followed by a voice: “Myna, are you here?”

“...Yes,” she answered weakly and softly.

“We’re all looking for you. May I come in?”

“...Hold on,” she said, moving away from the door onto the bed, wiping her tears, then saying, “Come in.”

He entered and immediately noticed something was wrong. After looking around, he asked gently, “Do you want to talk about it?”

Whether it was exhaustion or her actual desire to talk, she didn’t know, but she soon wept again, fighting to contain it, stopping only when she realized it was too late. Before she knew it, Arlen was at her side, comforting her with an arm around her.

"It's been so hard!" she sobbed.

"I know, Myna, I know," he replied.

"No, not just these battles over the past ten years. I can't remember a time when I wasn't worried or afraid of doing something wrong! Yet even after a decade of fighting, I still haven't…"

"I... I understand. Even after all the work you've put in, results don't always show immediately, but that's no reason to lose hope."

"I'm just so tired," she said more calmly. Then she looked into Arlen's eyes. For a moment, they simply stared at each other, their faces speaking more than a thousand words. As they slowly gravitated closer, loud knocking on the door came, followed by a yell, "Miss, are you there? The commander needs to speak to you!"

Mynedrie and Arlen awkwardly looked away from each other.

"I think we better get going," she said, rising and walking to the door while Arlen remained seated, unsure what to say.

They walked through the stone walls to the command room where the commander was speaking to two privates. Arlen recognized them—they were the men he had ordered to watch the lighthouse top in case Drumir tried something.

"Finally! Do you know how long I waited for you!" the commander said with frustration and disappointment.

Mynedrie braced for a long, humiliating lecture, only to be surprised when Arlen warmly greeted the two privates, "There you guys are! It's been too long!"

"Well, it's only been less than a day, really," one replied.

"And hadn't you ordered us to keep watch on the light tower at all times with your telescope?" the other added.

"And I assume you saw something interesting if you abandoned your post?" Arlen asked.

"Well, we saw corrupted militiamen put up a large metallic rod with a thin copper wire wrapped around it. We have no clue why," one explained.

"That's why we needed you two here quickly. You're our leading experts in Drumir and science, respectively," the commander added, still furious at the wait.

"Well, given what we know about what Drumir did with Stria, he might be trying something similar. What that pillar's for, I'm not sure," Arlen said.

"Me neither. Until now, Drumir preferred diseases and curses. Seems his muses have changed recently; he hasn't made anything like this before," Mynedrie added.

"Well, whatever it is, it'll likely be big. He must be desperate to turn the tide, and given how much steel, iron, bronze, and other resources he's stockpiled, I doubt my troops can take it on in their state. A charge on the castle is out of the question. If only we could put men on the lighthouse and destroy whatever that is," the commander said.

For a moment there was silence until Arlen realized, "Maybe we can!"

"What do you mean?" Mynedrie asked.

"Come with me and I'll show you," he answered.

Arlen, Mynedrie, the commander, and the two soldiers went to his home to the shed in his backyard. Arlen pulled out a large cloth so big he nearly couldn't hold it without it touching the ground, even folded.

He carefully laid it on the ground and unfolded it for the four onlookers.

When finished, he took out a massive metal bowl which he attached to the underside of the cloth where a small hole was placed.

"And what is this supposed to represent exactly?" the commander asked, puzzled.

"I call it a hot air balloon," Arlen said, catching his breath as setting everything up made him sweat.

"And what is this supposed to be?" the commander continued.

"This! The concept's simple. Remember those fire lanterns we send flying into the air each year toward the sea?"

"Yes," the commander replied.

"Well, this is that, but on a whole new level. According to my calculations, this one can carry up to two people."

"That seems low," Mynedrie observed.

"I know, but it's enough. One person pumps air in or partially seals it with this plate, and the other throws hooks to steer it—or else you're at the mercy of the wind," Arlen explained.

The commander looked concerned but said, "It's a long shot, but the only shot we have. Make your preparations, Mister Arlen. You and Miss Kianistra are going to fly!"

Chapter 30

All preparations had been met. The balloon was fully inflated with hot air and was tied down with a rope attached to the metal fence of Arlen's garden. Another rope led from the balloon's makeshift standing platform to a small group of heavily armored militiamen. Their job was to pull the balloon toward the castle as fast and as close as possible to prevent it from being carried away by the wind.

From there, Mynedrie would shoot an arrow with a rope attached toward the light tower, trying to aim at anything an arrow— even one being dragged down by the rope—could get stuck in. From there, they would try to attach the two hooks to the rails at the top of the light tower. Then they would have to destroy Drumir's device and get away before any corrupted militia could reach the top in time.

With one soldier loosening the rope that kept the balloon in place, they immediately started to rise, both Mynedrie and Arlen grabbing onto a rope that kept the platform attached to the balloon. Almost immediately after the balloon's rope was detached from the fence, soldiers downstairs started pulling them in the direction of the castle, not far from Arlen's house. They quickly made it to the city's eastern wall where the men on the ground were bombarded with arrows. Their armor and shields protected them, but it was not easy for Mynedrie and Arlen either. While the planks they stood on were thick enough to stop arrows, the platform was not big enough to fully protect the balloon. Luckily, shooting arrows nearly straight upward would quickly slow them down, though it still caused concern. Much of the archers' fire was countered by archers and crossbowmen on their side, shooting from rooftops and even from the closest parts of the city wall, causing arrows to fly in all directions.

Luckily, they were already around a hundred meters up, and most bows and crossbows were small-sized, so their arrows couldn't reach the balloon that high.

As they neared the top of the tower, almost close enough to

reach it with their grappling hooks, they saw Drumir standing at the edge with a wide, almost unnatural grin (whatever could pass as natural for something like him). Both Arlen and Mynedric looked on with dread, and Arlen expressed their shared feeling, “I don’t think we’re gonna like this!”

Right after, Drumir walked out of sight, and moments later, dark clouds started to gather over the light tower. They both realized what was coming next.

Arlen acted quickly, closing the coal seals and stopping air to the balloon to let it start descending. Yet, the balloon was still far above the tower tip where clouds gathered. The inevitable was coming fast, and he asked Mynedrie to shoot her specially made hook arrow through a window placed a few meters down the staircase.

Mynedrie took aim with a massive wooden bow, something she hadn’t used in a long time. Its draw was definitely heavier than her magical bow and took nearly all her normal arm's strength to pull. She felt her muscles strain, fighting the urge to shake while carefully aiming.

As the arrow shot off, carrying the rope, Mynedrie felt relief at releasing the bow. The arrow flew toward its target, breaking the glass. After it remained steady, they carefully checked it before they pulled as hard as they could to bring the balloon as close to the window as possible.

Direct swinging was not an option, as it would break at least a few bones. Instead, they crept closer until a sudden loud roar of lightning struck the balloon top, burning a hole through which hot air escaped, making the balloon crash downward in a burning mess of wood, balloon, and rope.

Mynedrie and Arlen did not fall with it, as they instinctively held on to the rope. The rope swung them violently against the wall. The impact would have broken both their bones if not for Arlen, who wrapped himself around Mynedrie—whose elven bones were more fragile—protecting her.

CRACK. Mynedrie heard multiple cracks and realized the

unsettling sound was Arlen, who couldn't hide that he'd broken several ribs. Yet when she looked into his eyes, she saw not only pain but relief and perhaps more.

Neither had time to reflect; neither had the energy to hang tens of meters above the ground, covered by a burning balloon emitting smoke carried away by the wind before it could reach them.

With every passing moment, it became harder to stay on the rope, so they tried climbing upwards, using the uneven, deep, and worn joints between the tower's stones. Arlen led, climbing as fast as pain and broken ribs allowed—a snail's pace. Usually, this would frustrate Mynedrie, used to climbing since a young age and finding humans' relative lack of nimbleness frustrating, but given the circumstances, she understood and encouraged him, "Come on, Arlen! Just a few meters! Keep going—you'll be there before you know it! Come on!"

Her encouragement gave Arlen strength to keep climbing

despite the pain. Before long, he reached the upper floor, carefully pushing away glass shards from the window frame without cutting himself. With Mynedrie's little push, he climbed through and turned around to offer her a hand.

When Mynedrie was safely inside, Arlen sat on stairs beside the dark, damp inner wall beneath candlelight.

For a moment, he wondered why the staff still performed basic maintenance despite their corruption by Drumir. Did Drumir design his crystals that way, or was it the person behind the crystal? He was pulled from thought by worsening pain as the excitement wore off.

Mynedrie squatted beside him, asking, "Arlen, are you alright? Where does it hurt?"

"Right rib cage," he answered, breathing heavily. "Bottle in my left pocket… give it… please…"

Mynedrie retrieved the flask and handed it to him. With weak, tired arms, he opened the bottle and took a sip. She carefully lifted his clothes to reveal his wounds. They were bad! Slowly, his wounds began healing, becoming less severe until the process stopped. The bones were back in place but still severely fractured, internal bleeding stopped, so he was likely to survive.

His rib fractures remained, meaning he was out of the fight.

"What!? How? Your wounds healed almost immediately?" Mynedrie asked.

"The divine elixir. It has healing properties. I convinced the remaining priests to give us a few bottles for our battle against Drumir. There are two others too—take them; you might need them."

"But your wounds, they're…"

"They'll be fine. Now go!"

Mynedrie sat beside him in silence, wanting to say more but not knowing what.

The silence broke as Arlen said, "Go show him what you're made of!"

"I will!" she said confidently and rushed upstairs, having

placed the remaining divine elixir bottles in her belt pouch.

She reached the door at the small hallway's end, atop the light tower. Behind it, she heard roaring thunder from Drumir's clouds—his plan was near completion.

Noticing the slightly open door and a cable slithering down the staircase, she understood it was likely attached to the metal device behind the door.

Taking her bow, Heart-Nailer, she stood ready and kicked open the door, drawing the bowstring, ready to shoot Drumir on sight. But he was nowhere, nor was the device.

Turning, she saw the cable moving skyward toward the flying device above the tower roof. She then heard sounds from the roof—Drumir, standing still, apparently guiding the machine with necromancy, excited each time his machine was struck by lightning. This was bad for Mynedrie.

She considered cutting the cable with a knife but realized cutting a cable carrying lightning was too dangerous.

Instead, she readied a light arrow with as much explosive

yield as possible. Moving from beneath the roof before Drumir reacted, she shot. The arrow flew, exploded on impact, producing a roar and bright light illuminating the tower top.

Drumir remained standing, hands before his face, eyes closed. Opening them after the thunder dimmed, he smiled, "You missed!" His smile turned to astonishment upon seeing metal debris and cable on the ground.

"I did not miss!" Mynedrie replied.

Drumir looked at her angrily, then raised his right hand and snapped his fingers.

Mynedrie was confused—not understanding the terrible thing he had just awakened. Beneath the courtyard, veiled under a blanket in his tent, something awakened, shining and oozing green light even through the cloth. Its limbs came alive with screeches of grinding metal.

Its short, squat legs raised its big, ash-dark, cylindrical body taller than the tent, lifting it to stand. From beneath the raised tent came a long metal arm with an extra joint, effectively a second elbow. At the end was a two-digit claw that grabbed the tent cloth and ripped it away, revealing itself completely.

“What is that thing?!” Mynedrie asked, unnerved.

“My creation—a weapon made of thick metal housing a large necromantic power source, powered by lightning. Your little stunt prevented me from fully charging it, but it still has enough power to largely complete its task.”

“Which is?” Mynedrie asked, afraid.

Drumir took from his robes a dark red sphere for communication and vision from the weapon, which held a similar bright sphere surrounded by dark steel armor.

“Why name it? No one of you will be left to remember it after it’s done anyway!” Drumir said, barely hiding a creepy smirk under his oversized grey-green nose. “Destroy all but the light tower!” he commanded the orb.

Immediately, the towering machine moved, its squatting legs bringing it to the front gate. It raised another smaller cylindrical shape, channeling green energy before unleashing a beam that pulverized the gatehouse. The stone construction stood no chance against the raw magical might. Arlen, hearing the screeches and debris raining from the destroyed gate tower, glimpsed the weapon and screamed in anger and disbelief: "My cannon... HE IS USING MY CANNON!"

"No…" Mynedrie said, shocked at the weapon's destructive power.

"Impressive, right? Too bad he can't use it much. Even without using it, his energy reserves will probably be depleted before midday tomorrow, seeing as you destroyed my lightning catcher before it was finished!"

"NO!!!" she screamed in wrath, flashed by painful memories of her village.

She shot arrow after arrow that would hit Drumir if not for the dark green shield created by a talisman around his neck. Light arrows were absorbed by the shield, which appeared each time Mynedrie fired.

Before she could change tactics, Drumir said, “Now it’s my turn!” He threw a grenade, which she hit mid-throw, causing it to explode in the air, spreading a deep dark fog.

Drumir glided from the fog, attacking with long metal nails she knew were poisoned. Dodging his first two slashes, she blocked the third with a knife, keeping her bow in the other hand.

He was surprisingly agile for an old, wrinkled, hunchbacked man, a trait rarely seen in the past ten years as his horrendous creation fought for him.

She tried to increase distance, freeing her hand by putting her bow over her shoulder.

This was a mistake, as Drumir shot green lightning bolts with incredible speed she could not dodge.

The lightning stiffened all her muscles except her metallic arm.

Instinctively, she pulled the arm in front of her to absorb

some pain. Instead, it fully absorbed the lightning streams, surprising both her and Drumir.

She advanced while Drumir fell back, using normal-sized nails again, releasing a poisonous snake-like creature from beneath his cloak. She sliced it mid-air, but its green blood—highly acidic—splattered on her face and clothing, biting into both. Worse, a second snake bit her right leg, puncturing her leather armor and injecting venom.

She killed the creature but was too late to stop the venom spreading.

She fell to her knees, pain stabbing her chest with erratic heartbeats slowing gradually. Drumir approached, triumphant, watching her struggle to breathe. "So close. Yet nothing you can do as you rot from the inside!" he said, smiling.

Placing a hand on her, he continued, "All we have to do is wait!"

"I can still stomp you into the ground…" she gasped.

"Please, my dear. Even unpoisoned, you could never hurt me. My amulet protects me, and your flesh would rot the moment you touch me. But don't worry. After your passing, I'll use one of my crystals to turn you into a weapon that, after all the headaches you gave me, will provide so much fun!"

Mynedrie's mind raced with horror about the pain he'd inflict postmortem, making her a pawn in his twisted game. For a moment, she felt defeated—until realization struck. She moved her left arm swiftly to grab his amulet and ripped it away with all her strength.

Drumir barely noticed his magical amulet had been stolen before he felt his left hand stabbed by one of Mynedrie's knives, emitting a possessed animal's shriek.

She looked into his watering eyes, holding his amulet in her leather-metal palm before crushing it with her augmented grip strength.

Seeing one of his most powerful items shattered made Drumir lose control. He tried to retreat, but Mynedrie threw her other knife, which stuck in his back.

Despite the pain, he moved toward the edge, intending to fly away on oversized wings he'd made. But a light arrow pierced his cloak and materials inside, hitting his weak torso. More arrows followed.

Drumir spread his right hand forward, desperately trying to grab escape but failed. Soon, his centuries-old necromancer body began falling apart, many parts turning to dust. His bones glowed green where charred remains stayed.

With Drumir gone, Mynedrie quickly drank two bottles of divine elixir as if her life depended on it—because it did.

She felt better; the chest pain vanished, burns from electricity and acid wounds disappeared, along with the bite punctures.

Moving to the balcony edge, she saw Drumir's giant metal monster continuing its rampage, tearing buildings and slaughtering innocents like vermin. It would destroy the city before running out of energy.

"No, not again. Never again!" Mynedrie said, her voice burning with defiance as she readied her bow.

She charged every ounce of energy into the bolt, making it brighter and bigger, forcing herself to push just a little more power, though she knew it might do little against the hulking monstrosity.

Then, strangely, a glowing string from her metal arm touched her bowstring, siphoning energy from her arm's crystal.

She was uncertain what was happening. The bolt changed shape—more streamlined, sharper, compact, and far more powerful.

When no more energy could be siphoned, she released the bolt. It flew faster than any arrow she'd shot, with unnatural precision.

To her surprise, the massive explosion she expected did not come. Instead, a wave of blue lightning washed over the iron beast, killing its energy source. The metal monster's joints screeched in desperation.

It became a statue, toppled by imbalance, now a monument to the ruins it helped create.

Mynedrie caught herself on the rail with her right arm, her left numb from the crystal's drained energy.

Ensuring she wouldn't fall, she whispered, "It's over… finally, it's over!"

Arlen came by, holding his side where his ribs had not fully healed. Seeing Mynedrie holding the rail, he moved to help, fighting pain, and asked, "Mynedrie, are you alright?"

"Yes, just need another crystal for this arm… and a good night's sleep," she said with a smirk, still watching the dead metal monster and running her thumb over her bow resting on the rail.

"Your bow?" Arlen asked.

"Broken, I know…" she said, her smirk turning to a sad frown. "It did what it needed to do. I can get over it. I should move forward," she said, though unsure.

"I guess we all should," he said.

She rested her head on his shoulder. Arlen leaned toward her, their eyes meeting in silent communication. Before either knew it, their lips moved closer, sharing a kiss as the sun rose to herald a new day.

Epilogue

And so, with Drumir slain and his creations either missing or destroyed, and the corrupted militia freed from their magical bondage, peace returned to the city of Drun. While the city suffered massive damage and both militia and populace endured thousands of casualties—with several hundred more wounded both majorly and minorly—the horrors of war, though scarring and traumatizing the city severely, did not succeed in killing it.

With Drumir's death, Drun would become a tourist attraction and somewhat of a pilgrimage site. The metal behemoth would be transformed into a monument attracting people of high social standing, especially elven aristocracy who had suffered massive financial losses every time Drumir set out to terrorize a place.

The elven aristocracy and their retinues, to a lesser degree, often spent small fortunes on anything considered related to what scholars would come to revere as 'Drumir's last trick.'

It was not just the nation's economy that transformed; Mynedrie saw a great shift in her life as well. Now that she could finally leave her decade-long hunt behind, she decided it was time for changes. She moved in with Arlen, now holding the noble title of 'Inventor' after his nickname, being given small plots of land, a mine, and the noble title of Earl by the new king—King Arnoudas 'the Bold' of House Drun. King Arnoudas was the brother of the last king of Drun, who, alongside many subjects cursed by the slain Drumir, perished in battle. Though more interested in the battlefield and a connoisseur of the finer things in life, King Arnoudas was a decent steward for the city and kingdom, making good use of opportunities to improve and expand trade wherever he saw fit.

His trip with Mynedrie and Arlen to the elven realm was a prime example, insisting the pair travel with him to ensure he had the ear of the richest and most influential elves.

While there, Mynedrie introduced her fiancé to her family, proud of her but somewhat controversial in her choice. Her mother had little good to say about Arlen. Her oldest sisters showed more enthusiasm: Aisyane with genuine happiness and relief that Mynedrie had safely returned home despite injuries; Yaismane shared similar relief but with a sense of improper curiosity; her older brother Daimir, the third child, was nowhere to be seen, and Mynedrie's questions about him were met with vague, uncomfortable answers. The younger twins bombarded Arlen with excited curiosity.

They were all saddened that Mynedrie had decided to move to Drun, each with differing degrees of understanding.

Her human wedding ceremony in a foreign temple almost caused her mother to storm out in rage. She chose not to attend, nor allow the twins to relocate. Only her two sisters and their families showed up.

As for Stria, no one had found her body or any sign of her existence except for a few rumors, some ridiculous, others unbelievable.

No one would find the soul crystal containing Sophia's soul trapped inside, nor were they aware her soul remained bound there. Most assumed she had passed and her body was petrified and placed inside the temple as a symbol of martyrdom, an example to the faithful.

The end

Made in United States
North Haven, CT
04 November 2026

21771530R00266